LIPIDS IN HEALTH: THEIR SOURCE, COMPOSITION, AND USES

LIPIDS IN HEALTH: THEIR SOURCE, COMPOSITION, AND USES

Edited by:

Shiv Sanjeevi and Prerna Pandey

Delve
PUBLISHING

www.delvepublishing.com

Lipids in Health: Their Source, Composition, and Uses

Shiv Sanjeevi and Prerna Pandey

Delve Publishing
2010 Winston Park Drive,
2nd Floor
Oakville, ON L6H 5R7
Canada
www.delvepublishing.com
Tel: 001-289-291-7705
001-905-616-2116
Fax: 001-289-291-7601
Email: orders@arclereducation.com

ISBN: 978-1-77407-340-7 (Hardcover)

DECLARATION

ABOUT THE EDITORS

Shiv Sanjeevi Sripathi secured second rank in Mumbai University and first rank in his college at his undergrad level and in his college during his Masters and has worked on a stem cell project at the Specialized Centre for Cell Based Therapy (SCCT), KEM Hospital at Mumbai, following which he was a Junior Research Fellow at Microbiology & Cell Biology Department, Indian Institute of Science, Bangalore from August 2009 to November 2011. The field of work involved cloning of cell wall genes and transcription factors in E. coli & M. smegmatis. This was followed by a stint as a Research and Development Executive at Transasia Biomedicals, Mumbai where he worked on patent writing and Immunoassays from 2011 till 2013. In 2013, he entered the teaching field where he has been tutoring biology and chemistry to the students of 11th and 12th grades along with English grammar and communication skills to people of all age groups.

Dr Prerna Pandey, a biotechnologist with several years of wet lab research experience, has worked as a senior research fellow at International Center for Genetic Engineering and Biotechnology, New Delhi. Her field of research in PhD included isolation and molecular characterization of Geminiviruses, genome sequencing, gene annotation, as well as gene silencing using the RNA interference technology. She has published papers in peer-reviewed journals in the field and has a submitted a number of annotated Geminiviral genome sequences in the GenBank, including two novel ones. Prerna has also completed her editing and proof-reading courses from SfEP, UK and now works as a freelance scientific editor, writer and a subject matter expert. She has edited and co-authored several books on a range of topics in Life-Sciences. When Prerna is not busy with her assignments, she loves visiting historical places.

TABLE OF CONTENTS

LIST OF CONTRIBUTORS

Shiv Sanjeevi Sripathi

Secured second rank in Mumbai University and first rank in his college at his undergrad level and in his college during his Masters and has worked on a stem cell project at the Specialized Centre for Cell Based Therapy (SCCT), KEM Hospital at Mumbai, following which he was a Junior Research Fellow at Microbiology & Cell Biology Department, Indian Institute of Science, Bangalore from August 2009 to November 2011. The field of work involved cloning of cell wall genes and transcription factors in E. coli & M. smegmatis. This was followed by a stint as a Research and Development Executive at Transasia Biomedicals, Mumbai where he worked on patent writing and Immunoassays from 2011 till 2013. In 2013, he entered the teaching field where he has been tutoring biology and chemistry to the students of 11th and 12th grades along with English grammar and communication skills to people of all age groups. Shiv also loves helping people write scientific manuscripts and has co-edited several books on life sciences and aspects of biotechnology with Apple Academic Press.

Dr Prerna Pandey

A biotechnologist with several years of wet lab research experience, has worked as a senior research fellow at International Center for Genetic Engineering and Biotechnology, New Delhi. Her field of research in PhD included isolation and molecular characterization of Geminiviruses, genome sequencing, gene annotation, as well as gene silencing using the RNA interference technology. She has published papers in peer-reviewed journals in the field and has a submitted a number of annotated Geminiviral genome sequences in the GenBank, including two novel ones. Prerna has also completed her editing and proof-reading courses from SfEP, UK and now works as a freelance scientific editor, writer and a subject matter expert. She has edited and co-authored several books on a range of topics in Life-Sciences. When Prerna is not busy with her assignments, she loves visiting historical places.

Amélie Bélanger-Lamonde

Département des Sciences Animales, Pavillon Paul-Comtois, Université Laval, Québec, QC, Canada G1K 7P4

Pallab K. Sarker

Département des Sciences Animales, Pavillon Paul-Comtois, Université Laval, Québec, QC, Canada G1K 7P4

Pierre Ayotte
Centre de Toxicologie, Institut National de Santé Publique du Québec (INSPQ), Québec, QC, Canada G1V 5B3
Axe Santé des Populations et Pratiques Optimales en Santé, Centre de Recherche du CHU de Québec, QC, Canada G1V 5B3

Janice L.Bailey
Département des Sciences Animales, Pavillon Paul-Comtois, Université Laval, Québec, QC, Canada G1K 7P4

Dominique P. Bureau
Fish Nutrition Research Laboratory, Department of Animal and Poultry Science, University of Guelph, Guelph, ON, Canada N1G 2W1

P. Yvan Chouinard
Département des Sciences Animales, Pavillon Paul-Comtois, Université Laval, Québec, QC, Canada G1K 7P4

Éric **Dewailly**
Axe Santé des Populations et Pratiques Optimales en Santé, Centre de Recherche du CHU de Québec, QC, Canada G1V 5B3

AlainLeblanc
Centre de Toxicologie, Institut National de Santé Publique du Québec (INSPQ), Québec, QC, Canada G1V 5B3

Jean-Philippe Weber
Centre de Toxicologie, Institut National de Santé Publique du Québec (INSPQ), Québec, QC, Canada G1V 5B3

Grant W. Vandenberg
Département des Sciences Animales, Pavillon Paul-Comtois, Université Laval, Québec, QC, Canada G1K 7P4

Hadis Sabour
Brain and Spinal Injury Research Center (BASIR), Tehran University of Medical Sciences, Tehran, Iran

Abbas Norouzi Javidan
Brain and Spinal Injury Research Center (BASIR), Tehran University of Medical Sciences, Tehran, Iran

Sahar Latifi
Brain and Spinal Injury Research Center (BASIR), Tehran University of Medical Sciences, Tehran, Iran

Mohammad Reza Hadian
Brain and Spinal Injury Research Center (BASIR), Tehran University of Medical Sciences, Tehran, Iran

Seyed-Hassan Emami Razavi
Brain and Spinal Injury Research Center (BASIR), Tehran University of Medical Sciences, Tehran, Iran

Farzad Shidfar
Nutrition Department, Iran University of Medical Sciences, Tehran, Iran

Mohammad Reza Vafa
Nutrition Department, Iran University of Medical Sciences, Tehran, Iran

Hamidreza Aghaei Meybodi
Endocrinology and Metabolism Research Institute (EMRI), Tehran University of Medical Sciences, Tehran, Iran

Ayssar A. Elamin
Department of Research and Development, LIONEX Diagnostics and Therapeutics GmbH, 38126 Braunschweig, Germany

Matthias Stehr
Department of Research and Development, LIONEX Diagnostics and Therapeutics GmbH, 38126 Braunschweig, Germany

Mahavir Singh
Department of Research and Development, LIONEX Diagnostics and Therapeutics GmbH, 38126 Braunschweig, Germany
Department of Genome Analytics, Helmholtz Centre for Infection Research, Inhoffenstrasse 7, 38124 Braunschweig, Germany

Yanfei Liu
Graduate School of Beijing University of Chinese Medicine, Beijing 100029, China
Institute of Clinical Pharmacology, Xiyuan Hospital of China Academy of Chinese Medical Sciences, Beijing 100091, China

Mingyue Sun
Institute of Clinical Pharmacology, Xiyuan Hospital of China Academy of Chinese Medical Sciences, Beijing 100091, China

Hezhi Yao
Institute of Clinical Pharmacology, Xiyuan Hospital of China Academy of Chinese Medical Sciences, Beijing 100091, China

Yue Liu
Cardiovascular Diseases Center, Xiyuan Hospital of China Academy of Chinese Medical Sciences, Beijing 100091, China

Rui Gao
Institute of Clinical Pharmacology, Xiyuan Hospital of China Academy of Chinese Medical Sciences, Beijing 100091, China

M. Snel
Departments of Endocrinology & Metabolism and General Internal Medicine, Leiden University Medical Center, 2300 RC Leiden, The Netherlands

J. T. Jonker
Departments of Endocrinology & Metabolism and General Internal Medicine, Leiden University Medical Center, 2300 RC Leiden, The Netherlands

J. Schoones
Walaeus Library, Leiden University Medical Center, 2300 RC Leiden, The Netherlands

H. Lamb
Department of Radiology, Leiden University Medical Center, 2300 RC Leiden, The Netherlands

A. de Roos
Department of Radiology, Leiden University Medical Center, 2300 RC Leiden, The Netherlands

H. Pijl
Departments of Endocrinology & Metabolism and General Internal Medicine, Leiden University Medical Center, 2300 RC Leiden, The Netherlands

J. W. A. Smit
Departments of Endocrinology & Metabolism and General Internal Medicine, Leiden University Medical Center, 2300 RC Leiden, The Netherlands

A. E. Meinders
Departments of Endocrinology & Metabolism and General Internal Medicine, Leiden University Medical Center, 2300 RC Leiden, The Netherlands

I. M. Jazet
Departments of Endocrinology & Metabolism and General Internal Medicine, Leiden University Medical Center, 2300 RC Leiden, The Netherlands

LIST OF ABBREVIATIONS

AARS	Alma Aquaculture Research Station
AST	Aspartate aminotransferase
BMI	Body mass index
BALP	Bone alkaline phosphatase
BMD	Bone mineral density
BALP	Bone-specific alkaline phosphatase
BASIR	Brain and Spinal Injury Research Center
CSCs	Cancer stem cells
CVD	Cardiovascular disease
CNS	Central nervous system
CKD	Chronic kidney disease
CBEs	Cocoa butter equivalents
CV	Coefficient of variation
CHD	Coronary heart disease
DAG	Diacylglycerol
DIO	Diet-induced obesity
DHA	Docosahexaenoic acid
DXA	Dual-energy X-ray absorptiometry
EPA	Eicosapentaenoic acid
EMRC	Endocrinology and Metabolism Research Center
ER	Endoplasmic reticulum
EFAs	Essential fatty acids
FPG	Fasting plasma glucose
FATP	Fatty acid transport protein
FCR	Feed conversion ratio
FFA	Free fatty acid

HCC	Hepatocellular carcinoma
HDL	High density lipoprotein
HFD	High-fat diet
HOMA-IR	Homeostatic model assessment-insulin resistance
HSL	Hormone-sensitive lipase
HGL	Human gastric lipase
HPL	Human pancreatic lipase
IGT	Impaired glucose tolerant
IRS	Insulin-receptor substrates
IDL	Intermediate-density lipoprotein
IUNS	International Union of Nutritional Sciences
IUPAC	International Union of Pure and Applied Chemists
LCAT	Lecithin-cholesterol acyl-transferase
LDs	Lipid droplets
LCT	Long-chain triacylglycerols
LDL	Low density lipoprotein
MHC	Major histocompatibility complex
MALDI	Matrix-assisted laser desorption ionization
MCFA	Medium-Chain Fatty Acids
MFGM	Milk fat globule membrane
NGT	Normal glucose tolerant
PPAR-α	Peroxisome proliferator-activated receptor alpha
POPs	persistent organic pollutants
PC	Phosphatidylcholines
PUFA	Polyunsaturated fatty acid
RCTs	Randomized controlled trials
ROS	Reactive oxygen species
SIMS	Secondary ion mass spectrometry
SCFA	Short-chain fatty acids
SCI	Spinal cord injury
SREBP	Sterol regulatory element-binding protein
TFAs	Trans fatty acids

TAGs	Triacylglycerols
TCA	Tricarboxylic acid cycle
TNF	Tumor necrosis factor
UPR	Unfolded protein response
VLCDs	Very low-calorie diets
VLDL	Very low-density lipoproteins

PREFACE

"This is a modified version of the old diet. The DASH diet was a real breakthrough for lowering blood pressure and we changed it. We reduced the carbohydrate content and replaced it with unsaturated fat or protein, and it lowered blood pressure more and improved lipids, and overall cardiovascular risk goes down"- Frank Sacks.

"Biological energy comes from the sun. Light energy harvested by photosynthesis in chloroplasts and phototropic bacteria becomes stored in carbohydrates and fats. This stored energy can be released by oxidative metabolism in the form of adenosine triphosphate (ATP) and used as fuel for other biological processes."- John E. Walker

The above are quotes from history about lipids. These heterogeneous molecules are a component of our diet. Research now points out that completely avoiding lipids from the diet can have serious consequences. This book serves to address new facets of the aspects of these biomolecules. With a plethora of information available on the internet and diet books, there is a requirement for data available along with sound scientific proof that has been provided in this book.

For instance, this book discusses how rather than the total fat consumed, one should focus on the relative ratio of lipids or unsaturated fatty acids consumed. Similarly, research shows that it is advised to analyze changes in the levels of individual lipid species rather than the entire lipid profile in disease, as such alterations hold great potential to understand the mechanism of disease as well as treatment.

Chapter 1 discusses the classification and importance of lipids. Chapter 2 discusses the role of lipids and food and how we should look at the balance in what fats are eaten rather than what is eaten. Chapter 3 then summarizes the role of lipids in health ranging from the brain to the skin and hair. Finally, Chapter 4 explains the role of lipids in disease. Several case studies have been presented to describe the alterations reported in lipid molecules in various diseases to hence, allow lipids to be used for diagnosis and prognosis. This becomes important given the paucity of reliable and stable biomarkers currently. Finally, the use of lipids as drug delivery agents has also been described.

This is one of the first short books that describe the role of lipids in food, disease and health using scientific evidence. It is hoped that the reader(s) can obtain all the important data, research and outlooks in this concise yet deep presentation.

Editors

CHAPTER 1

WHAT ARE LIPIDS: AN INTRODUCTION

Shiv Sanjeevi Sripathi, Dr Prerna Pandey

[1]Shiv Sanjeevi Sripathi secured second rank in Mumbai University and first rank in his college at his undergrad level and in his college during his Masters and has worked on a stem cell project at the Specialized Centre for Cell Based Therapy (SCCT), KEM Hospital at Mumbai, following which he was a Junior Research Fellow at Microbiology & Cell Biology Department, Indian Institute of Science, Bangalore from August 2009 to November 2011. The field of work involved cloning of cell wall genes and transcription factors in E. coli & M. smegmatis. This was followed by a stint as a Research and Development Executive at Transasia Biomedicals, Mumbai where he worked on patent writing and Immunoassays from 2011 till 2013. In 2013, he entered the teaching field where he has been tutoring biology and chemistry to the students of 11th and 12th grades along with English grammar and communication skills to people of all age groups. Shiv also loves helping people write scientific manuscripts and has co-edited several books on life sciences and aspects of biotechnology with Apple Academic Press.

[2]Dr Prerna Pandey, a biotechnologist with several years of wet lab research experience, has worked as a senior research fellow at International Center for Genetic Engineering and Biotechnology, New Delhi. Her field of research in PhD included isolation and molecular characterization of Geminiviruses, genome sequencing, gene annotation, as well as gene silencing using the RNA interference technology. She has published papers in peer-reviewed journals in the field and has a submitted a number of annotated Geminiviral genome sequences in the GenBank, including two novel ones. Prerna has also completed her editing and proof-reading courses from SfEP, UK and now works as a freelance scientific editor, writer and a subject matter expert. She has edited and co-authored several books on a range of topics in Life-Sciences. When Prerna is not busy with her assignments, she loves visiting historical places.

1.1 INTRODUCTION: LOOKING AT FATS IN THE DIET

The concept of "What is a healthy diet?" is challenging to be answered by many including clinicians. The amount of data available in the media, research papers and researchers display varying results while several of the interpretations are reported as misinterpretations and many are contradictory! While the functioning of nutrients in lowering the risks of diseases is yet to be firmly established, there are reports of adopting patterns that involve the incorporation of legumes, whole-grain foods, vegetables, and fruits while lowering the intake of full-fat dairy products, added sugar containing foods, red meat, and refined starches. Health requires an amalgamation of regular exercise, maintaining weight, avoiding smoking and a proper diet (Skerrett and Willett, 2010).

The typical Western diet has been linked to several disorders such as cardiovascular conditions and atherosclerosis; this diet is described as containing sugar, highly processed grains, and red meat while the amount of fiber, whole grains, fruits and vegetables is a lack! The U.S. Department of Agriculture (USDA) lists the Dietary Guidelines for Americans that "*provide authoritative advice for people two years and older about how good dietary habits can promote health and reduce risk for major chronic diseases.*" These guidelines were presented as the Food Guide Pyramid that however lacked distinction in certain nutritional components that was hence, replaced with the abstract MyPyramid shown below:

Figure 1.1: http://www.mypyramid.gov/global_nav/media_resources.html.

"Permission to Reprint: All of the informational materials produced by the USDA Center for Nutrition Policy and Promotion, whether printed or maintained electronically on this website, are in the public domain and as such are not restricted by copyright law unless otherwise stated. We ask only that informational materials, both graphics and text, provided by CNPP be reproduced as originally designed and/or written and that they not be altered or edited in any way. For accuracy and continuity of the message, we encourage all users to reproduce the information as original designed and/or written"-USDA

The stripes shown in the pyramid show various food groups and their relative extents in a healthy diet and are color coded from left to right as orange for grains, green for vegetables, red for fruits, yellow for oils, blue for dairy and purple for meat and beans.

Another scheme has been provided by the Department of Nutrition at Harvard School of Public Health called the Healthy Eating Pyramid,

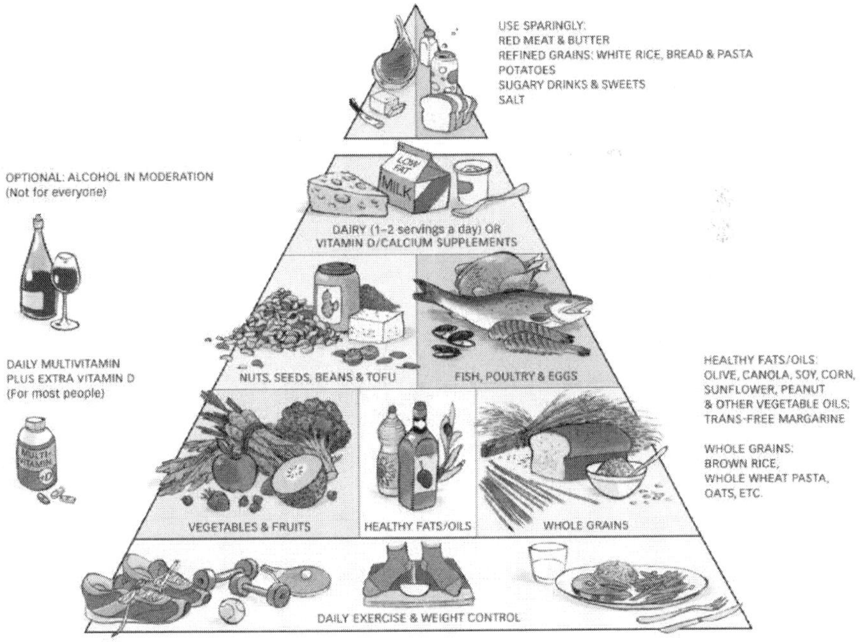

Figure 1.2: http://www.hsph.harvard.edu/nutritionsource/what-should-you-eat/pyramid/index.html

Permission to reprint: It belongs to the President and Fellows of Harvard College. *"It may be used without permission for educational and other non-commercial uses with proper attribution, including the following copyright notification and credit line: Copyright © 2008. For more information about The Healthy Eating Pyramid, please see The Nutrition Source, Department of Nutrition, Harvard School of Public Health, http://www.thenutrition-source.org, and Eat, Drink, and Be Healthy, by Walter C. Willett, M.D. and Patrick J. Skerrett (2005), Free Press/Simon & Schuster Inc."*

Among the aspects of diet are dietary fat or lipids. To give an introduction, the statement that "fat is bad" has been there since the 20th century. This statement, however, has been associated with poor diets that lack fats but offer calories and hence resulted in several diseases. Health depends on different types of fats: ***Trans fats*** increase low-density lipoprotein (LDL) cholesterol levels while lowering high-density lipoprotein (HDL) cholesterol that is associated with damage to the arteries and health. ***Saturated fats*** increase both LDL and HDL and are sourced from dairy products and red meat; with a recommendation of 8% of the intake every day. ***Monounsaturated*** and ***polyunsaturated*** fats are sourced from fish and nuts that are associated with stabilized heart rhythms and lowered LDL and higher HDL. Overall, while the exact proportion of fats in the diet still requires characterization; low amounts of trans and saturated fat and more levels of unsaturated fats are found to lower the risk of diabetes and cardiovascular disease (Skerrett and Willett, 2010).

1.2 CLASSIFYING LIPIDS

A loose definition of lipids is a set of organic compounds that dissolve in water but are not soluble in water (Smith, 2000). Another definition given to facilitate classification of lipids is that they are "hydrophobic or amphipathic small molecules that may originate entirely or in part by carbanion- based condensations of thioesters (fatty acids, polyketides, etc.) and/or by carbocation-based condensations of isoprene units (prenols, sterols, etc.)" (Fahy et al, 2005). Structurally and functionally, these are a heterogeneous group of molecules that cause considerable variations in the classification systems of these molecules. The sites Lipid Library and Cyberlipids divide these molecules into simple lipids that yield 2 groups of molecules when hydrolyzed while complex lipids yield 3 or more groups. Another group of derived lipids is described by LipidBank. The International Lipid Classification and Nomenclature Committee has described a comprehensive system to classify lipids using chemical and biochemical principles (Fahy et al, 2011).

Fahy et al, 2005 presented a comprehensive system of classifying lipids based on the various hydrophobic and hydrophilic elements that make up a lipid.

Table 1.1: The categories of lipid classification are summarized below

Category and Abbreviation	**Example**
Fatty acyls: FA	dodecanoic acid
Glycerolipids: GL	1-hexadecanoyl-2-(9 *Z*-octadecenoyl)-*sn* -glycerol
Glycerophospholipids: GP	1-hexadecanoyl-2-(9 Z-octadecenoyl)- sn -glycero-3-phosphocholine
Sphingolipids: SP	*N*-(tetradecanoyl)-sphing-4-enine
Sterol lipids: SL	cholest-5-en-3-ol
Prenol lipids: PR	2*E* ,6*E* -farnesol
Saccharolipids: SL	UDP-3- *O*-(3*R*-hydroxy-tetradecanoyl)-d-*N*-acetylglucosamine
Polyketides: PK	aflatoxin B1

This system lacks the group of glycolipids to focus on lipid structures and glycosylated lipids have been placed on the basis of their core. While protein modification by lipids is observed, this classification does not include these here.

- The fatty acyls (FA) are fundamental categories of biological lipids. The hydrophobic nature of this group is due to repeating series of methylene groups. Examples: straight-chain saturated fatty acids, cyclic fatty acids, octadecanoids and lipids in the jasmonic acid pathway, prostaglandins, leukotrienes, docosanoids, fatty acid esters, Long-chain ethers.
- The glycerolipids (GL) are glycerol-containing lipids. Examples: mono-, di- and tri-substituted glycerols including the fatty acid esters of glycerol (acylglycerols).
- Glycerophospholipids (GP) are separate from the above as they are abundant and are components of membranes and give rise to messengers. Several classes are seen on the basis of nature of the polar "head group" at *sn* -1 position for archaebacteria or the *sn*-

3 position of eukaryotes and eubacteria.

- Sphingolipids (SP) contain a serine- fatty acyl-CoA derived sphingoid base backbone. They have several classes such as sphingoid bases and their simple derivatives, ceramides, phosphosphingolipids.
- Sterol lipids (ST) are divided using biological function while the number of carbons in the core skeleton is used to classify the steroid families. C18 steroids comprise of estrogen family while androgens comprise the C19 steroids.
- Prenol lipids (PR) are produced from isopentenyl diphosphate and dimethylallyl diphosphate (mevalonic acid pathway). Examples include Vitamin A, quinones and hydroquinones, Vitamins E and K.
- Saccharolipids (SL) are described as fatty acids linked directly to a sugar backbone as against glycolipids that contain the fatty acyl portion of the molecule is present in a glycosidic linkage: International Union of Pure and Applied Chemists (IUPAC). Examples are acylated glucosamine precursors lipid A of Gram-negative bacteria, Nod factors of nitrogen-fixing bacteria, acylated trehalose units.
- Polyketides (PK) are a diverse set of molecules. Several drugs have been derived from this group of molecules such as tetracyclines, avermectins, and erythromycins (Fahy et al, 2005).

1.3 LIPIDS IN FOODS

Crude or processed foods yield dietary lipids that have several structures. 97% of the dietary fatty acids or fats are Triacylglycerols (TAGs) with polar lipids (phospholipids) following next (Armand, 2008). There are several classes of acylglycerol based on the glycerol backbone, the bonds such as ester bonds or ether bonds. Such a diversity of lipids also is augmented when food habits are altered or newer sources of food are introduced.

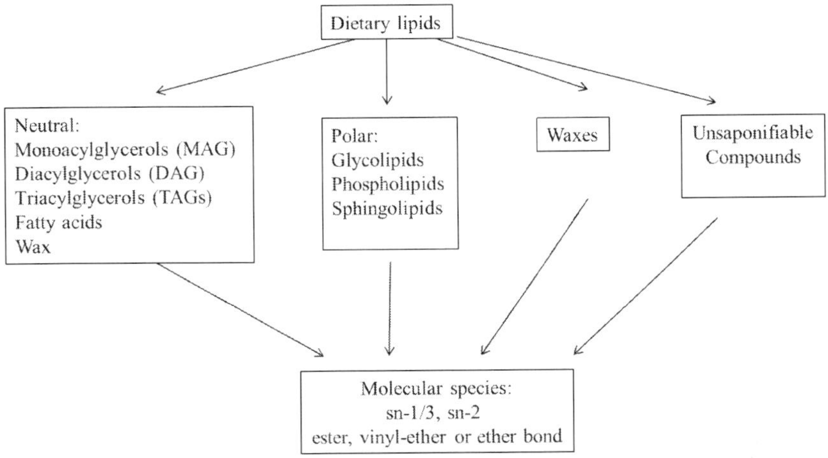

Figure 1.3: The chief dietary lipids and structural diversity.

Among various food materials, there is a variation supramolecular structure of lipids. There are oleosomes in oil seeds, supramolecular organizations in milk fat globules. There are TAG cores with an interface or a membrane that has a composition, organization. For example, the TAG core involves the stabilization of a monolayer interface while in raw milk has a phospho- and sphingo-lipids, cholesterol and protein membrane.

(a) Egg yolk

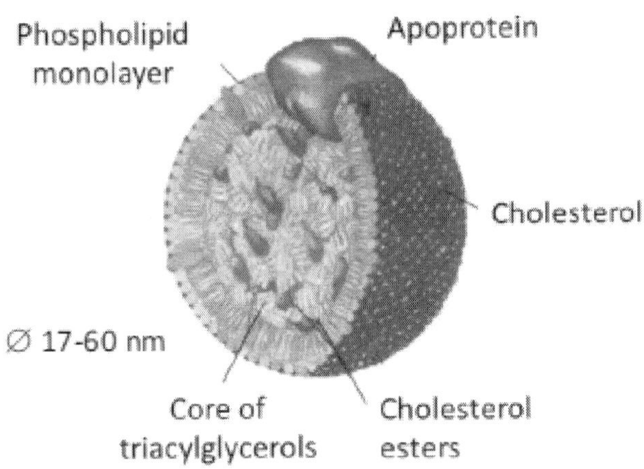

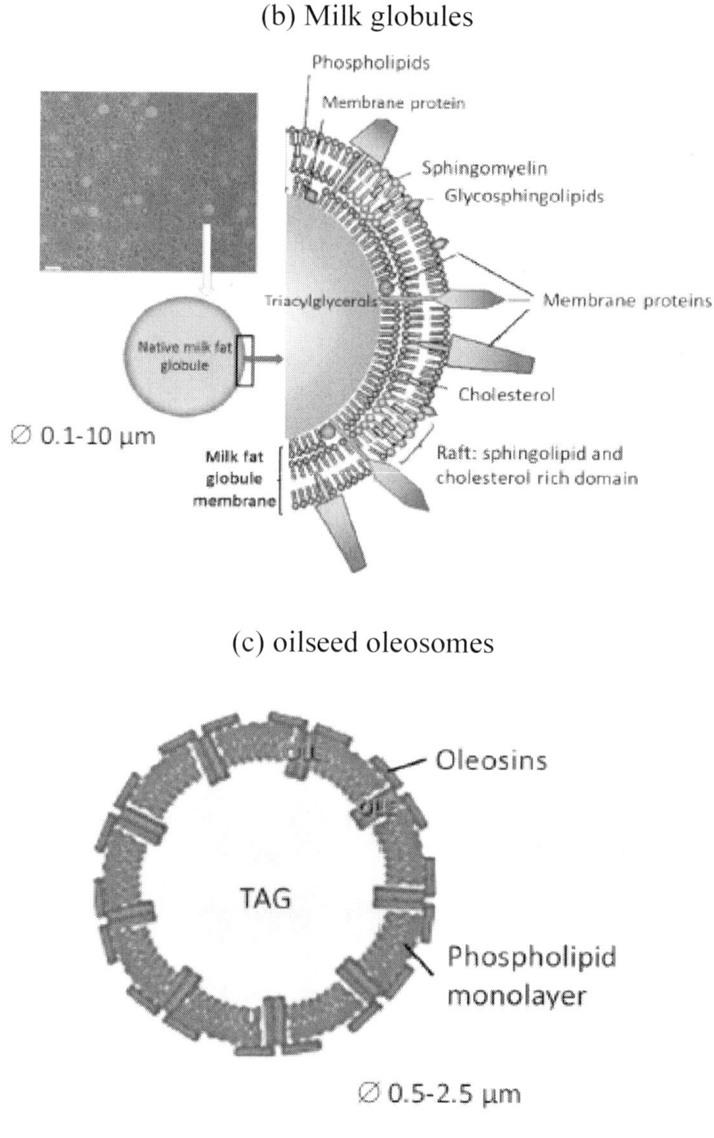

Figure 1.4: Various structures of native structures of lipids in foods (Meynier and Genot, 2017).

There are lipid organizations in formulated foods such as water-in-oil and oil-in-water emulsions. The fate of various dietary fatty acids is influenced by lipid structures in the Gastro Intestinal tract of humans as below:

In the mouth chewing begins the process of digestion, the digestion and absorption of lipids is initiated by feeling the food in the mouth. While rats possess lingual lipase to release free fatty acid (FFA) while this enzyme has not yet been detected in humans. The human gastric lipase (HGL) of the stomach hydrolyzes TAGs that is essential to:

- Allow emulsification by the formation of surface-active molecules
- Short and medium chain Fas are used for energy
- To condition and allow cholecystokinin (CCK) to stimulate pancreatic lipase

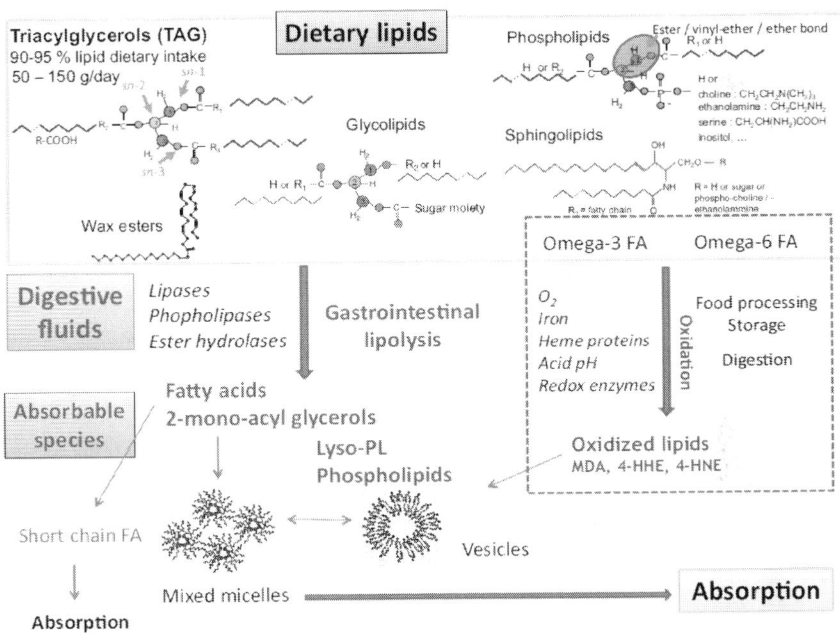

Figure 1.5: Various absorbable lipids (Meynier and Genot, 2017).

The lipolysis is influenced by the organization of lipids in a meal. Lipolysis was more when native milk was used against milk formula. Most of the lipolysis is carried out by human pancreatic lipase (HPL) whose secretion was more in a diet with a high-fat diet (45–50 g lipids) against a low-fat one (12.5 g) in healthy subjects. The availability of fatty acids and hence metabolism is influenced by the structure of lipids in food.

1.3.1 Digestion of Lipids for Health

Human health can be augmented by the manipulation of lipid bioavailability in foods. This can achieve the optimization or increase of low concentration yet beneficial lipids (for example omega-3 PUFA for the general public). Another scenario is the lowering of lipids absorbed for patients with metabolic issues or weight issues.

- The aggregates of TAGs at a molecular level and colloidal organizations of lipid in foods at the supramolecular level are being pursued.
- The use of catalysts or lipases can modify the TAG structure of natural fats and oils to modify their properties. For example, a blend of lipids from vegetables called Betapol shows TAGs similar to those in human milk. Such a structure allows lowered fecal loss and hence increased absorption of palmitic acid
- For fat malabsorption subjects, the modification of TAGs to contain PUFA (polyunsaturated fatty acids) in sn-2 and medium-chain FA in sn-1 and sn-3 positions can allow efficient absorption.
- Stabilizing agents and emulsifiers in the food industry involve the use of lecithins from egg or vegetables or krill. Depending on their source, the lecithin shows changes in composition. The polar lipids in the lecithin bind the TAG surface and lower the susceptibility to destabilization as well as interacting with other components of the food.
- Novel foods that provide sensory and health benefits can involve the introduction of antioxidants or vitamins.
- The control of the oil and water phases can allow for a component to be released in a specific site such as the tissue or any portion of the GI tract.

(Meynier and Genot, 2017)

Hence, the design of lipids in food can allow for several benefits in food:

- Bioavailability
- Taste
- Appearance
- Texture
- Stability
- Metabolism

- Price
- Balance
- Consumer acceptance

(Meynier and Genot, 2017)

The following

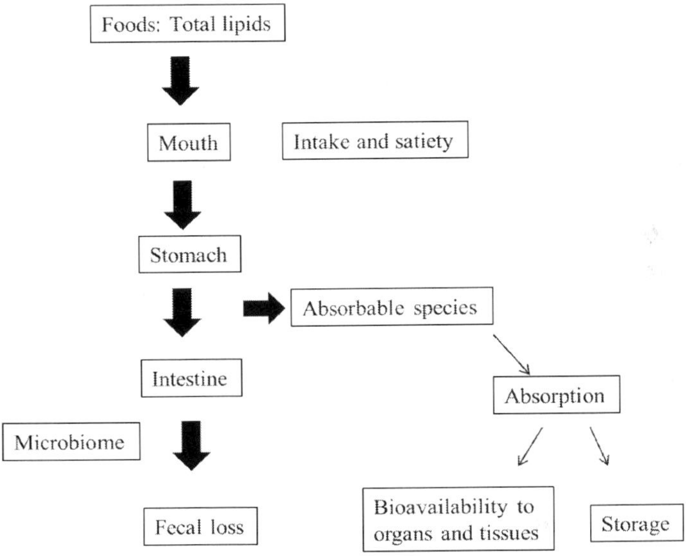

Figure 1.6: shows a summary of lipids in our food and our body to hence control health.

(Adapted from Meynier and Genot, 2017)

1.4 AGE, HEALTH AND LIPIDS

Age is reported as a major and an important risk factor in health with changes at a genetic level controlling aging, for example, p66shc adaptor protein controls this process (Camici et al, 2015). This and other such proteins are involved in modulating lipids as well as other cardiovascular risk factors. While treatment approaches for anti-aging are being pursued, until their realization, the role of age is vital in order to predict the risks of developing diseases. According to the ESC Guidelines for the Management of Dyslipidaemias, high lipoprotein (a) concentrations have been associated with cardiovascular disease. While treatments have looked at addressing the levels of LDL, the lowering of the lipoprotein has not seen active studies.

Several approaches that have been shown as promising are:

- Lipid apheresis of lipoprotein(a)
- phase I and II trials have shown a 90% reduction of lipoprotein using antisense therapy targeting apolipoprotein(a)
- Inhibitors to PCSK9 (proprotein convertase subtilisin/kexin type 9) decrease the levels of lipoprotein and LDL (Lu¨scher, 2017).

The migration of monocytes into walls of arteries has been shown to play a role in the inflammation of arteries or atherosclerosis. Apart from lowering the level of LDL to address the inflammation, research shows the use of monoclonal antibodies against PCSK9. A study published in 2017 studied the effect of these antibodies in familial hypercholesterolemia (22 patients) who were not exposed to statins along with controls. They found that the patients showed an increased expression of chemokine receptor (CCR) 2 (3-fold) as compared with controls that were associated with higher plasma LDL-cholesterol levels, intracellular lipid accumulation and higher migration of monocytes. The administration of PCSK9 monoclonal antibody for 24 weeks showed lowering of plasma LDL by 49% along with CCR2 and monocyte migration (Moens et al, 2017).

There is a loss of benefits offered by HDL-cholesterol in several cardiovascular diseases. A study published in 2017 examined the effect of HDL and mortality in the LURIC study of 3310 coronary angiography patients. Mortality was predicted by the levels of dimethylarginine in the serum with higher levels of the molecule in HDL linked to higher mortality. The elevated amounts of HDL with low symmetric dimethylarginine were linked to lessened mortality (Zewinger et al, 2017).

1.5 FINAL POINTS

This chapter presented an introduction to lipids in the diet from a different perspective as presented by media. A comprehensive system of classification was also presented that can allow for studying the new field of lipidomics. The processing of lipid molecules in various parts of digestion and modification of lipids in foods was discussed to allow for desirable features and mainly an augmentation of health. Finally, health depends on the age factor that has been targeted since the time of fairy tales! 2 case studies were presented that show a link between lipid profiles, age and hence, cardiovascular function and health.

1.6 REFERENCES

1. Armand M. Digestibilité des matières grasses chez l'homme. *Sci Alim.* 2008; 28:84–98.
2. Bernelot Moens SJ, Neele AE, Kroon J, van der Valk FM, Van den Bossche J, Hoeksema MA, Hoogeveen RM, Schnitzler JG, Baccara-Dinet MT, Manvelian G, de Winther MP, Stroes ES. Proprotein convertase subtilisin/kexin type 9 monoclonal antibodies reverse the pro-inflammatory profile of monocytes in familial hypercholesterolaemia. *Eur Heart J.* 2017; 38:1584–1593.
3. Camici GG, Savarese G, Akhmedov A, Luscher TF. Molecular mechanism of endothelial and vascular aging: implications for cardiovascular disease. *Eur Heart J.* 2015; 36:3392–3403.
4. Fahy E, Cotter S, Sud M, Subramaniam S. Lipid classification, structures and tools. *Biochim Biophys Acta.* 2011 November; 1811(11):637–647.
5. Fahy E, Subramaniam S, Brown HA, et al. A comprehensive classification system for lipids. *J Lipid Res.* 2005; 46:839-861.
6. Lu¨scher TF. Prevention: further insights on age and lipids. *Eur Heart J.* 2017; 38:1525–1529.
7. Meynier A, Genot C. Molecular and structural organization of lipids in foods: their fate during digestion and impact in nutrition. *Oilseeds & fats Crops and Lipids.* 2017; 24(2):D202.
8. Skerrett PJ, Willett WC. Essentials of healthy eating: a guide. *J Midwifery Womens Health.* 2010;55(6):492-501.
9. Smith, AD. Oxford Dictionary of Biochemistry and Molecular Biology. Oxford University Press; Oxford: 2000. Rev
10. Zewinger S, Kleber ME, Rohrer L, et al. Symmetric dimethylarginine, high-density lipoproteins and cardiovascular disease. *Eur Heart J.* 2017; 38:1597–1607.

CHAPTER 2

LIPIDS IN FOOD AND HEALTH

Shiv Sanjeevi Sripathi[1], Dr Prerna Pandey[2]

[1]Shiv Sanjeevi Sripathi secured second rank in Mumbai University and first rank in his college at his undergrad level and in his college during his Masters and has worked on a stem cell project at the Specialized Centre for Cell Based Therapy (SCCT), KEM Hospital at Mumbai, following which he was a Junior Research Fellow at Microbiology & Cell Biology Department, Indian Institute of Science, Bangalore from August 2009 to November 2011. The field of work involved cloning of cell wall genes and transcription factors in E. coli & M. smegmatis. This was followed by a stint as a Research and Development Executive at Transasia Biomedicals, Mumbai where he worked on patent writing and Immunoassays from 2011 till 2013. In 2013, he entered the teaching field where he has been tutoring biology and chemistry to the students of 11th and 12th grades along with English grammar and communication skills to people of all age groups. Shiv also loves helping people write scientific manuscripts and has co-edited several books on life sciences and aspects of biotechnology with Apple Academic Press.

[2]Dr Prerna Pandey, a biotechnologist with several years of wet lab research experience, has worked as a senior research fellow at International Center for Genetic Engineering and Biotechnology, New Delhi. Her field of research in PhD included isolation and molecular characterization of Geminiviruses, genome sequencing, gene annotation, as well as gene silencing using the RNA interference technology. She has published papers in peer-reviewed journals in the field and has a submitted a number of annotated Geminiviral genome sequences in the GenBank, including two novel ones. Prerna has also completed her editing and proof-reading courses from SfEP, UK and now works as a freelance scientific editor, writer and a subject matter expert. She has edited and co-authored several books on a range of topics in Life-Sciences. When Prerna is not busy with her assignments, she loves visiting historical places.

2.1 LIPIDS AND FOOD

A host of food items contain fats and oils as their contents that are involved in several aspects (Charley & Weaver, 1998):

- Aerate butter
- Increase the tenderness of cakes
- Lubricate foods
- Add flavor to foods
- Introduce a moist feeling in the mouth
- They influence the texture of cakes
- The order of release of elements in a food item is influenced by fat
- The transfer of heat to a food item is done via fats.
- The texture, feeling in the mouth, ability to spread and stability of food is influenced by fat crystals melting in food.
- The functions of pies, pasta, breads are influenced by fats shortenings (Rios et al, 2014).

The trend of overweight or obese population is on the rise; attributed to feeding trends and health. However, as fat has a major role in satiety, appearance, and texture of foods, it is challenging to address this aspect of optimally lowering the content of fats in foods efficiently. The risk of diseases is higher with the use of *Trans* fatty acids (TFAs) and saturated fats. Several replacers have been tested to replace fat foods.

The various roles of lipids in food are summarized in Figure 2.1:

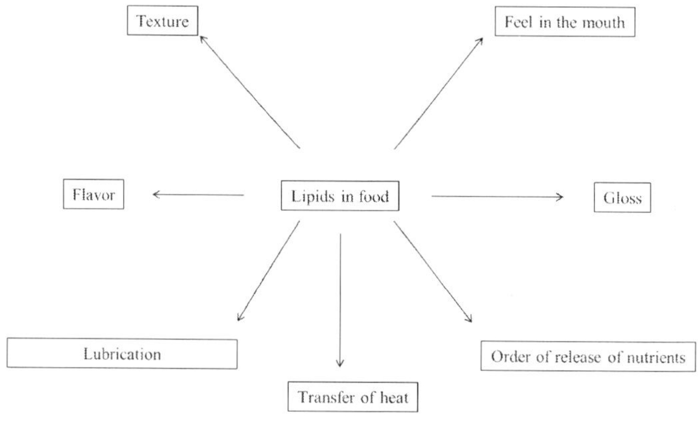

Figure 2.1

2.1.1 Lipids in the Bakery

The batter for cakes is an oil-in-water emulsion with fats dictating the technical aspect as well as sensory appeal. Fats allow for trapping air, softness, texture, feel and structure in cakes. Due to the presence of excess fats in cakes mainly *Trans* fatty acids (TFAs), research has attempted to lower the levels of the excess fat. A 2011 study involved the study of properties of cakes following the use of sunflower oil (SFO) and coconut oil (CNO) along with emulsifiers while hydrogenated vegetable fat was used for controls. The quality, viscosity, and content of lauric acid were higher in the test cakes as opposed to controls (Kumari et al, 2011).

The process of bread making involves the incorporation of 2 to 5 % fat in relation to flour; Fats in bread making allow for improved retention of gas, aeration, lubrication, and texture with several shortening agents added such as margarine and butter. Several formulations have been tested to replace the fat content of bread: fat mimetics refer to carbohydrate or protein substances that resemble fats in terms of functions and sensorial properties.

- Palm oil in sweet bread was replaced with emulsion monoglyceride + co-surfactant + water + sunflower oil or monoglyceride + co-surfactant + water + oil palm known as hydrogel caused a reduction of 81 % of saturated fat with other features similar to that of palm oil.
- In another study, inulin was used to replace sugar and lipid in scone quick bread that offers high-quality food albeit with lesser calories.
- While inulin and oligosaccharides replaced 10% fat and sugars in quick breads, the characters were similar and volume was increased in the case of the former.
- The use of Benefat® could substitute hydrogenated fat in hamburger buns.
- A 2013 study reported the use of inulin to replace fats in short dough biscuits; the use of inulin up to replace 20% of the fat yielded biscuits with similarity to controls (Rodríguez-García et al, 2013).

2.1.2 Lipids in Chocolate

Among the ingredients of chocolates, the lipids have the greatest bearing to control the release of flavors, melting, and gloss. The fats used are usually milk fat and cocoa butter. The latter is costly and is subject to various factors such as the weather that has led to a search for replacements of this component to decrease the level of fats. The introduction of cocoa butter equivalents (CBEs) allows the replacement of cocoa butter without affecting the texture.

- Coconut oil or palm kernel was found to be a replacement for cocoa butter in terms of flavor release, feel in the mouth and hardness.
- However, several replacers are available such as caprocaprylobehenic triglyceride or caprenin that was used in chocolates with lowered calories. Salatrim is another replacement used in chocolate-flavored coatings. Other examples are avicel -microcrystalline cellulose, olestra, inulin, and carrageenan.
- The use of β-glucan-rich hydrocolloid lowered the caloric value of chocolates and also produced a soft texture.

2.1.3 Lipids in Ice Creams

In the case of ice creams, lipids play a role in:

- Melting in the mouth
- Texture
- Flavor
- Color
- Retention of shape
- Creaminess

In order to lower the extent of fats used in ice creams, several alternatives have been explored such as:

- 20 % of skimmed milk powder could be substituted by soybean extract that showed resistance to melting and no major changes in sensory properties.
- The replacement of fat with green coconut pulp in chocolate ice cream showed similarity in properties as well as acceptance by a majority of tasters. A benefit to the use of the pulp is the presence of medium chain fatty acids.
- The use of 4-6% inulin to replace fat resulted in similar viscosity profiles, consistency, and textures to those of the controls (Rios et al, 2014).

2.2 LIPIDS AND FOOD INDUSTRY

Deterioration of food quality is caused mainly by oxidation of lipids that changes the flavor into rancidity as well as changes in nutrition and color. There can also be the formation of toxicity as well as a compromise in the shelf life that hence becomes a challenge in the industry. In the case of polyunsaturated fatty acids, the challenge is oxidative instability of beneficial polyunsaturated fatty acids in food.

There are several applications of lipids in food:

- Lipids help in the retention of carbon dioxide in the dough to augment the volume of dough.
- The sensory properties of foods are influenced by lipids.
- The saliva production is augmented that increases the viscosity.
- Flavors produced by plant lipids on account of increased unsaturated lipids.

Rancidity or oxidation of lipids is a major roadblock for oils and fats that are classified as a nutraceutical.

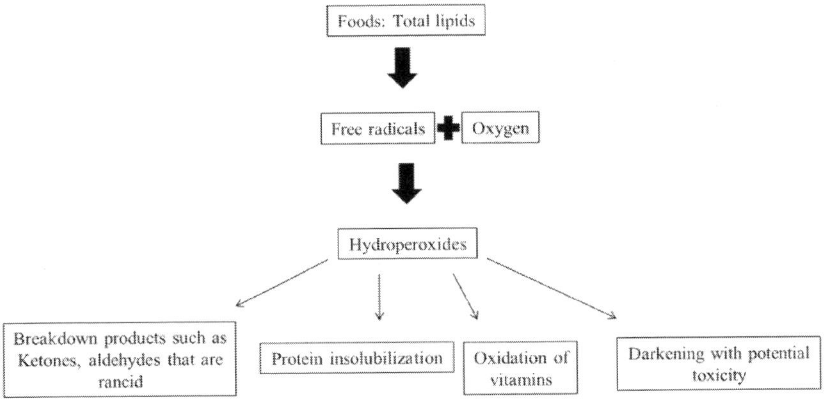

Figure 2.2: The process of oxidation involves the formation of several break-down products.

The process involves:

- Initiation: With the formation of free radicals.
- Propagation: With chain reactions.
- Termination: non-radical products that end the reactions.

The scheme is shown below in Figure 2.3:

$$\begin{aligned}
\text{Initiation} \quad & ROOH \leftrightarrow ROO\cdot + H \\
& ROOH \leftrightarrow RO\cdot + OH\cdot \qquad (1.1) \\
& 2ROOH \leftrightarrow RO\cdot + H_2O + ROO\cdot \\
& ROO\cdot + RH \leftrightarrow ROOH + R\cdot \qquad (1.2) \\
\text{Termination} \quad & ROO\cdot + ROO \leftrightarrow ROOR + O_2 \\
& RO\cdot + R \leftrightarrow ROR \qquad (1.3)
\end{aligned}$$

Figure 2.3

Mainly unsaturated fatty acid undergoes lipid oxidation with increased oxidation and increased. This rancidity leads to targeted functions, quality, shelf life, and taste. The unsaturated fatty acids contain double bonds that augment the susceptibility to rancidity. Such damages due to rancidity can be altered using antioxidants that augment the shelf life and maintain the quality. Those used in food processing should be economical, lack toxicity, and possess stability at lower levels. The antioxidant choices are reflected in the type of product and regulations.

Examples of antioxidants include:

Natural: Carotenoids, Ascorbic acid, phospholipids, sterols, Sulfur dioxide, Tocopherols

Artificial: Butylated hydroxytoluene, Butylated hydroxyanisole, Tocopherols, Ethoxyquin, Gallates

Hence, the food rancidity can be controlled by the level of lipid oxidation (Ali et al, 2016).

2.3 LIPIDS IN BUTTERMILK FOR HEALTH

The health benefits of milk fat globule membrane (MFGM) have received the attention of the research community. The lipids of this fraction (phospholipids) have been reported to lower cholesterol, check inflammation, and target neurodegeneration among others. A study published in 2014 reviewed the benefits of sweet buttermilk on health; here the buttermilk is referred to the aqueous fraction obtained from the churning of cream into butter. The mechanical destabilization of the oily phase or butterfat and the aqueous phase is caused by air and repeated physical collisions. The scheme is shown below in Figure 2.4:

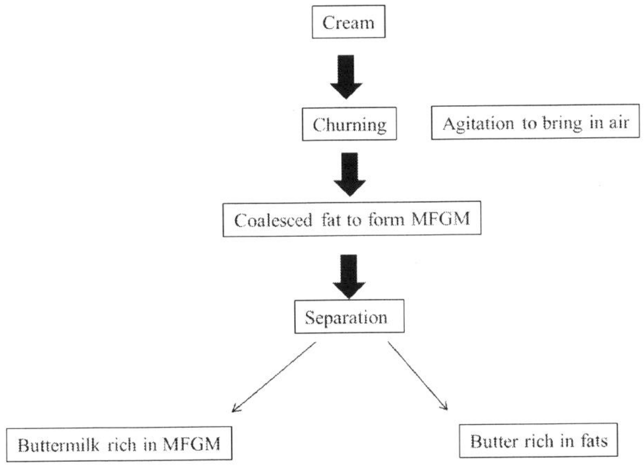

Figure 2.4

The buttermilk formed has a protein profile similar to that of skim milk (80%) while the fat contents are different. The various MFGM components in buttermilk bestow the health properties observed. This MFGM is made of:

- Polar lipids as a single internal layer
- Protein coat that is dense
- Polar lipids and proteins form an outer bilayer

The lipids of the buttermilk include 26 to 30% polar lipids that account for 60 to 70% of all the polar lipids of milk along with 70% neutral lipids. The polar lipids are glycerophospholipids and sphingolipids. Glycerophospholipids are made of a polar head mainly a phosphate linked to choline, serine, ethanolamine or inositol with a hydrophobic tail of C14–C24 fatty acid chains (mainly medium-to-long unsaturated fatty acids) attached to glycerol. A similar chain of fatty acids is attached to sphingosine in sphingolipids. Most sphingomyelins in MFGM involve saturated fatty acids; these MFGM sphingomyelins form lipid rafts by interacting with cholesterol as rigid domains. These rafts function in trafficking and cell signaling (Conway et al, 2014).

2.3.1 Lipids and their Health Benefits

The phospholipids of milk have been linked to benefits in diseases such as inflammation, cardiovascular disease, and cancer since the 20th century. The sphingolipid metabolites such as ceramides are messengers involved in angiogenesis, apoptosis, cell growth, and immune functions. The functions of MFGM are attributed to the weakened domains formed by the lipid rafts that allow the action of digestive enzymes or allowing signaling circuits (Conway et al, 2014)

The structure system is shown as a scheme in **Figure 2.5**:

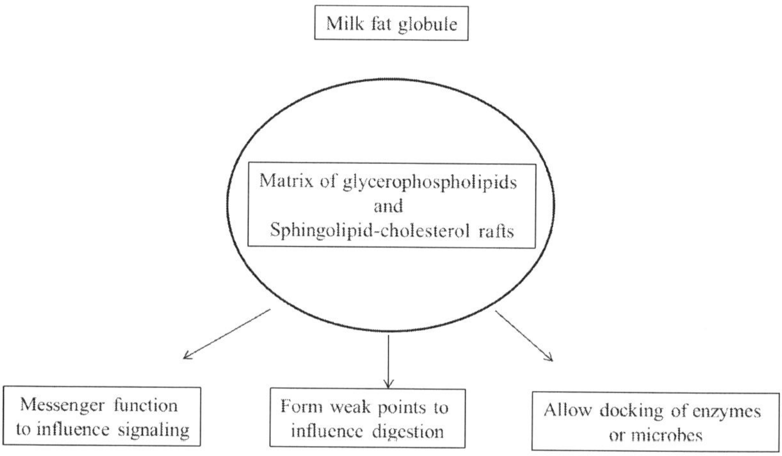

Figure 2.5

The following is a summary of several studies reported regarding the effects of lipids on health:

- In a randomized control trial, milk fortified with sphingomyelin promoted neurobehavioral development in infants born with very low weights. The sphingomyelin treatment was associated with the parameters of neurodevelopment (Tanaka et al, 2013).
- A 2012 study showed that the episodes of 3-day fevers and their duration in 182 healthy children was lowered by milk fat globule membrane (INPULSE) enriched formula milk. Additionally, positive changes in behavior were also reported as measured by the Achenbach System of Empirically Based Assessment that checks for depression, anxiety, cognition, aggression, etc (Veereman-Wauters et al, 2012).
- Endoplasmic reticulum stress is associated with neurodegenerative disorders; this stress was lowered in vitro by 8.4% sphingomyelin plus 8.3% phosphatidylethanolamine and 1.9% phosphatidylcholine (all milk phospholipids) (Nagai, 2012).
- A 2010 article reported the effect of MFGM from buttermilk ultrafiltration on cancer in a rat model. The administration of

MFGM from buttermilk with 0.11% sphingomyelin, 0.03% sphingomyelin in corn oil or anhydrous milk fat showed that the product from buttermilk could protect cells from cancer. The role of sphingomyelin in addressing apoptosis-resistant cells was suggested by the authors (Snow et al., 2010).

- In another work published in 2013, the proliferation of HT-29 colon cancer cells was inhibited by MFGM isolates from raw milk *in vitro* (Zanabria et al., 2013).
- A 2013 study showed that the use of MFGM concentrates from buttermilk was found to inhibit rotavirus infection of monkey kidney cells *in vitro*. Thus, the role of lipids to target infection was also shown (Fuller et al, 2013).
- A double-blind, randomized, placebo-controlled crossover study involving 34 healthy subjects with mild hypercholesterolemia showed the distinct reduction of plasma cholesterol and triglyceride by consumption of buttermilk. Patients consumed 45 g of a placebo with < 0.1%sphingomyelin or buttermilk with 0.6% sphingomyelin randomly but consecutively; each compound for 4 weeks and a switch to the other compound for 4 weeks. The level of LDL-cholesterol in serum was lowered by 5.6% compared with placebo. The mechanism was associated with sphingomyelin inhibiting cholesterol in the intestine (Conway et al, 2014).

The various health benefits from lab/clinical studies are shown as a summary in **Figure 2.6**:

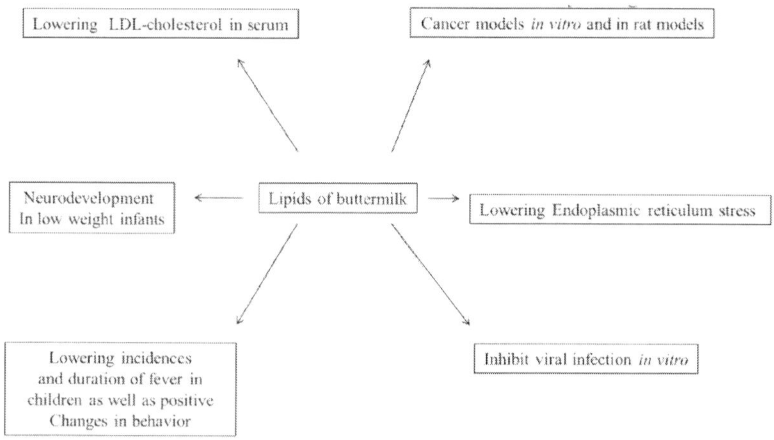

(Conway et al, 2014)

Figure 2.6

2.4 CASE STUDY OF LIPIDS AND HEALTH: DEEP FRIED FOODS

Among the culinary techniques, deep frying involves the use of a high-temperature fatty medium to attain a particular texture and taste. The combination of high temperatures, moisture, air and the food item all bring about changes in the oil or fat used for frying. Several molecules are altered by the action of the hot oil mainly the hydroperoxides that are formed when lipid degradation begins and are further degraded into other products. Thermo-oxidation of oils in some foods yields alkyl furans or polycyclic aromatic hydrocarbons that are carcinogens. Apart from these chemical transformations, there are changes in the physical nature too such as browning of potatoes. Another challenge is the presence of *trans*-fatty acids in the oils used for frying with evidence that these fatty acids are associated with cardiovascular diseases.

A study published in 2018 examined the quality and health status of deep-frying processes in the absence of supervision or regulation as seen with vendors. The rationale behind this study was an increase in the extent of obese or overweight individuals in Chile attributed to alterations in the eating habits of the Chilean population. This country was reported to have the first rank for processed food consumption among Latin American countries. Another trend is the increase in the consumption of fast food from street vendors without any heed to quality. The food from these vendors was classified as fried that was associated with cardiovascular disease (the *numero uno* reason for mortality in Chile) in addition to concerns of sanitary protocols. These protocols refer to the open-air environments, food and oil types as well as reuse of oil. As the high school and university students are the main target population of these foods, such a study was performed to avoid health complications in this population.

The quality of oil in foods was examined by physiochemical trait studies of the most popular joints as determined by student questionnaires; the food items were French fries, cheese empanadas, vegetarian empanadas, spring rolls, sopapillas, and churros. The fat percentage, fatty acid composition, and oil quality parameters were studied. The authors reported that the fat content in these food items were 5.9% to 32.1% with contents of more than 20% in French fries, cheese empanadas, sopaipillas, and churros of which the first 2 were preferred by the students despite the risks on health.

Degradation of fatty acids by hydrolysis was suggested by the content of free fatty acids of 0.38% to 1.48%. Oxidation in fats was evaluated by peroxidation with a prescribed value of < 2meq O2/kg to be considered as good quality vegetable oil. The values in this study were 1.7meq O2/kg to 103.3meq O2/kg suggestive of more than 90% of the foods was of not good quality. Oxidation was also confirmed by assessments of K27 values. The deterioration of fats and oils was also confirmed by :

- High *p*-anisidine values that indicate the 2,4-dienal and 2-alkenyl oils aldehydes.
- All samples exceeded a prescribed TOTOX index of more than 4 (103.6–395.0).
- The polyene index or polyunsaturated-to-saturated fatty acid ratio was more than the prescribed value of more than 4 representing a deteriorated state of lipids.
- The ratio of linoleicto- palmitic fatty acids (i.e., C18:2-to-C16:0) was more than the set value of 2%.

The study presented the quality of the fat present in street foods (with more than 90% of university students as consumers) to study the extent of thermooxidative deterioration due to the frying process. The results showed "highly oxidized and hydrolyzed" fat present in the food article analyzed. This becomes important as the health risks associated with lipid deterioration come into the picture that calls for discard from sales.

(Flores et al, 2018)

2.5 DIETARY FAT AND DIET

The risk of coronary heart disease (CHD) with the consumption of saturated and unsaturated fat was discussed at the International Expert Movement to Improve Dietary Fat Quality in cooperation with the International Union of Nutritional Sciences (IUNS) symposium.

The salient points are:

- As far as the risk of dietary saturated fatty acids and heart disease is concerned it was recommended that dietary saturated fatty acids should be partially replaced with unsaturated fatty acids in order to lower the risk of CHD. Major limitations in studies conducted were highlighted such as the duration of the study and the ethnicity of the participants who are exposed to a particular kind of diet. A good guideline was to maintain 10% saturated fatty acids as it is not desirable and not possible to replace the entire 100% with unsaturated fatty acids.
- The risk of CHD was linked to the type of dietary fat more than the quantity of total fat taken. The risk of CHD is lower with consuming n-6 and n-3 PUFA and intake of whole grains rather than refined carbohydrates.
- The need for a general food guide was acknowledged over that of a nutrient-based one.
- Healthier fats involve the increased consumption of fat-free plain yogurt and plant oils over animal fats.
- The use of lesser refined or highly processed grains, foods with sweeteners and red meat was suggested.
- The authors overall acknowledged the idea that fats are not enemies but a balance to the types of fats is key in healthy eating (Nettleton et al, 2018).

2.6 FINAL POINTS

As this book delves into the importance of lipids in health, the idea that fats are not enemies were discussed with the help of research journals. Apart from the exact role of lipids in food and the search for fat replacers in the food industry, this chapter also looked at how to avoid oxidation of fats in processed foods to avoid health issues. The beneficial results of buttermilk lipids in various clinical and laboratory studies were presented that can serve as avenues of future research. A case study to highlight the quality of fat in popular fried foods along with health implications was also presented.

Finally, a symposium recommendation was given that rather than looking at the total amount of fats, the balance in what fats are eaten makes the difference.

2.7 REFERENCES

1. Ali AH, Abed SM, Korma SA, Hassan HM. Lipids: Functions, Applications in Food Industry and Oxidation. *Am J Food Sci Nutr Res.* 2016; 3(6):162-171.
2. Charley H, Weaver C. *Foods: a scientific approach* (3th ed.). New Jersey: Prentice-Hall; 1998.
3. Conway V, Gauthier SF, Pouliot Y. (2014) Buttermilk: Much more than a source of milk phospholipids. *Animal Frontiers*. 2014; 4(2): 44-51.
4. Flores M, Meyer L, Orellana S, et al. Quality of Lipid Fractions in Deep-Fried Foods from Street Vendors in Chile. *J Food Quality*. Volume 2018, Article ID 7878439, 8 pages.
5. Fuller KL, Kuhlenschmidt TB, Kuhlenschmidt MS, Jiménez-Flores R, Donovan SM. Milk fat globule membrane isolated from buttermilk or whey cream and their lipid components inhibit infectivity of rotavirus in vitro. *J Dairy Sci*. 2013; 96:3488–3497.
6. Kumari R, Jeyarani T, Soumya C, Indrani D. Use of vegetable oils, emulsifiers and hydrocolloids on rheological, fatty acid profile and quality characteristics of pound cake. *J Texture Studies.* 2011; 42:377-386.
7. Nagai K. Bovine milk phospholipid fraction protects Neuro2a cells from endoplasmic reticulum stress via PKC activation and autophagy. *J Biosci Bioeng*. 2012; 114:466–471.
8. Nettleton JA, Brouwer IA, Mensink RP. Fats in Foods: Current Evidence for Dietary Advice. *Ann Nutr Metab.* 2018; 72:248–254.
9. Rios RV, Pessanha MDF, de Almeida PF. Application of fats in some food products. *Food Sci Technol, Campinas*. Jan.-Mar. 2014; 34(1): 3-15.
10. Rodríguez-García J, Laguna L, Puig A, Salvador A, Hernando I. Effect of fat replacement by inulin on textural and structural properties of short dough biscuits. *Food Bioproc Tech.* 2013; 6:2739-2750. http://dx.doi.org/10.1007/s11947-012-0919-1
11. Snow DR, Jimenez-Flores R, Ward RE, Cambell J, Young MJ, Nemere I, Hintze KJ. Dietary milk fat globule membrane reduces the incidence of aberrant crypt foci in Fischer-344 rats. *J Agric Food Chem*. 2010; 58:2157–2163.
12. Tanaka K, Hosozawa M, Kudo N, Yoshikawa N, Hisata K, Shoji H, Shinohara K, Shimizu T. The pilot study: Sphingomyelin-fortified milk

has a positive association with the neurobehavioural development of very low birth weight infants during infancy, randomized control trial. *Brain Dev*. 2013; 35:45–52.

13. Veereman-Wauters GSS, Rombaut R, Dewettinck K, Deboutte D, Brummer R.- J, Boone M, Le Ruyet P. Milk fat globule membrane (INPULSE) enriched formula milk decreases febrile episodes and may improve behavioral regulation in young children. *Nutrition*. 2012; 28:749–752.
14. Zanabria R, Tellez AM, Griffiths M, Corredig M. Milk fat globule membrane isolate induces apoptosis in HT-29 human colon cancer cells. *Food Funct*. 2013; 4:222–230.

CHAPTER 3

LIPIDS AND HEALTH (PHYSICAL AND MENTAL)

Shiv Sanjeevi Sripthi

Shiv Sanjeevi Sripathi secured second rank in Mumbai University and first rank in his college at his undergrad level and in his college during his Masters and has worked on a stem cell project at the Specialized Centre for Cell Based Therapy (SCCT), KEM Hospital at Mumbai, following which he was a Junior Research Fellow at Microbiology & Cell Biology Department, Indian Institute of Science, Bangalore from August 2009 to November 2011. The field of work involved cloning of cell wall genes and transcription factors in E. coli & M. smegmatis. This was followed by a stint as a Research and Development Executive at Transasia Biomedicals, Mumbai where he worked on patent writing and Immunoassays from 2011 till 2013. In 2013, he entered the teaching field where he has been tutoring biology and chemistry to the students of 11th and 12th grades along with English grammar and communication skills to people of all age groups. Shiv also loves helping people write scientific manuscripts and has co-edited several books on life sciences and aspects of biotechnology with Apple Academic Press.

3.1 INTRODUCTION

Lipids are involved in several roles such as their incorporation in organelles such as the plasma membrane, the Golgi apparatus, lysosomes among others. Lipids have been named the "Cinderellas" of cell biology on account of them being understood to a lesser degree than that of proteins. Several functions of lipids are associated with signaling and structural roles as well as their forms as glycolipids or farnesylation/ palmitoylation of proteins. There are lipid rafts on which research is focused as to their role. The majority of what is known is from the lab work on membranes. While the classification of proteins is based on their functions; that of lipids is on the basis of their chemical nature. A combination of imagining and MS shows promise to study the distribution of lipids; other techniques such as Secondary ion mass spectrometry (SIMS) and Matrix-assisted laser desorption ionization (MALDI) allows for visualization.

The visualization of lipids can involve the process of fluorescently tagging. The identification of lipid binding domains can be facilitated by the use of in situ synthesis of tagged lipids that involves reaction or clicking of a lipid with a labeled probe. Such clicked lipids can be observed and visualized within a cell. Due to the localization of cellular lipids into vesicles, the resolution of conventional fluorescence microscopy can be augmented by super-resolution microscopy. Research in the field of lipid biology holds much promise with using modern analytical approaches and membrane models (Muro et al, 2014).

3.2 HAIR BIOLOGY AND LIPIDS

Initial work using a mouse model with a knockout of peroxisome proliferator-activated receptor-γ (PPARγ) specific for the mammary gland showed that this mouse manifested no significant phenotype. However, the pups that fed on the milk with oxidized lipids of this mutant showed alopecia that was reversed when the pups were not administered this milk (Wan et al, 2007). Later work using gene array analysis and real-time PCR in human cicatricial alopecia revealed that patients with lichen planopilaris (LPP) showed a reduction in genes associated with lipid metabolism, synthesis of cholesterol and synthesis of peroxisomes. PPARγ-knockout in hair follicle stem cells of mouse showed a similar trend observed with LPP along with the formation of cysts and inflammation of the dermis (Karnik et al, 2009).

The involvement of cholesterol in hair has been a question since a long time. A 2010 study involved the knock out of the enzyme that is rate limiting in cholesterol synthesis: 3-hydroxy-3- methylglutaryl coenzyme A reductase using Insig proteins that inhibit the enzyme. This knockout showed accumulated sterol precursors in the follicle associated with distorted hair shafts and whiskers, distal follicles plugged by keratin as well as catagen/telogen forms that were arrested abnormally (Evers et al, 2010). The increased levels of sterols in the tissues were linked to alterations in Sonic Hedgehogs (Hh) pathway.

The involvement of lipids in the damages to hair follicles observed can be explained using several approaches:

- The work by Karnik et al, 2009 involved a suggestion of inactivated PPARγ that is associated with modulation of metabolism and inflammation. As this factor is involved in inflammation, its dysregulation is associated with damage to tissues such as the follicle.
- Metabolism of lipids occurs in organelles such as endoplasmic reticulum (ER), peroxisomes and mitochondria. An example is the sensing of cholesterol by the ER membrane. Hence, disturbances of lipid metabolism in the hair follicle can be associated with inflammation and damage.
- An alternative mechanism is a modification by lipids of signaling proteins involved in morphogenesis of hair like Wnts and Hedgehogs (Hhs).

Thus, the role of lipids in morphogenesis and cycling of the hair follicle is suggested with alterations in lipid metabolism associated with disorders of hair (Stenn and Karnik, 2010).

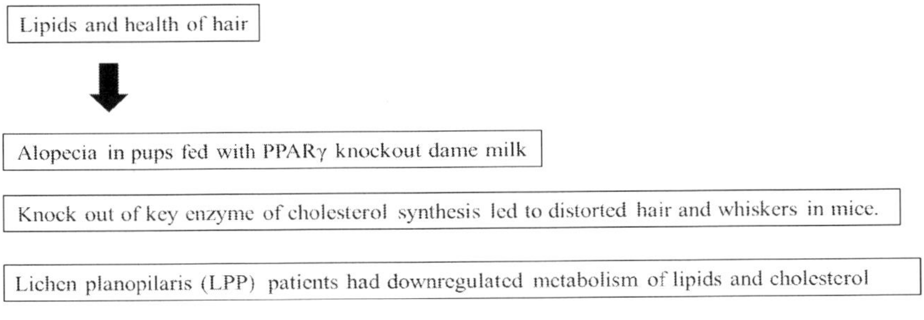

Figure 3.1: The summary of findings in lipids in the health of hair.

3.3 LIPIDS IN THE IMMUNE SYSTEM

The role of certain fatty acids in immunity has been shown by epidemiological studies as seen with lower levels of inflammation associated issues in Greenland Eskimos who consume long-chain *n*-3 polyunsaturated fatty acids in their diet. The role of fatty acids on the immune system involves several mechanisms such as:

- Alterations in phospholipids of the membrane
- Peroxidation of lipids
- Alteration of gene expression
- Production of eicosanoids
- Apoptosis regulation by docosahexaenoic acid [DHA], eicosapentaenoic acid [EPA], palmitic acid [PA] and arachidonic acid [AA].

The effect of free fatty acids or administration of lipids has been observed on several aspects of immune functions:

- Phagocytosis
- Proliferation of lymphocytes
- Cell surface molecule modifications
- Cytokine production
- Natural killer (NK) cell activities

The immune modulation is dependent on the nature and concentration of fatty acids as well as the duration of administration. Different fatty acids exert a different effect on modulating the immune system:

- Many functions are distinctly lowered by long-chain *n*-3 polyunsaturated fatty acids, including EPA (20:5*n*-3) or DHA (22:6*n*-3).
- The suppression of apoptosis by *n*-6 polyunsaturated fatty acids has been shown to play a chemo-promotive role in cancer.
- Olive oil contains mainly oleic acid that is an immune modulator.
- The administration of fatty acids (*n*-3 polyunsaturated fatty acids/ *n*-3 or *n*-9 fatty acids) was found to decrease the proliferation of lymphocytes.
- *n*-9 monounsaturated fatty acids or *n*-3 polyunsaturated fatty acids lower the expression of cytokine receptors, modify phagocytosis, inhibit NK cell action and inhibit antigen presentation.

3.3.1 How do they work?

The following hypotheses have been presented for the effect of lipids on immune system functions:

- The changes in fatty acids affect the fluidity of the membrane
- They are involved in altering transcription factors.
- They modify the production of eicosanoid.
- Lipid peroxides are produced.

The incorporation of long-chain *n*-3 polyunsaturated fatty acids, such as EPA lowers the production of IL-1 and tumor necrosis factor (TNF) inhibitor called prostaglandin E2. EPA also lowers translocation of NF-ḰB as well as altering the proliferation and subsets of lymphocytes. The expression of major histocompatibility complex (MHC) class II molecules on monocytes was altered by polyunsaturated fatty acids. Adhesion molecules of peripheral blood mononuclear cells were lowered when a diet containing monounsaturated fatty acids was consumed. The consumption of fish oil containing diet in animals showed a suppression of the proliferation of Th1 cells.

3.3.2 Lipids and Apoptosis

The regulation of functions involves the process of apoptosis to eliminate aberrant or damaged cells. Apart from the induction of apoptosis by radiation or chemicals such as ceramides or infectious agents such as bacteria or virus, fatty acids have also been involved. EPA and DHA (polyunsaturated fatty acids) or palmitic acid, as well as fish oil, has found to induce cell death involving the mitochondria. This involves an alteration of the mitochondrial transmembrane potential or activation of the cascade of caspases by the fatty acids.

The expression of Bcl-2 was lowered while that of Fas ligand increases with dietary fatty acids. The expression of Bcl-2 is lowered to cause apoptosis when polyunsaturated fatty acids are increased in the diet. The localization of Ras (involved in apoptosis) is lowered with fish oil. The induction of apoptosis also depends on lipid peroxidation.

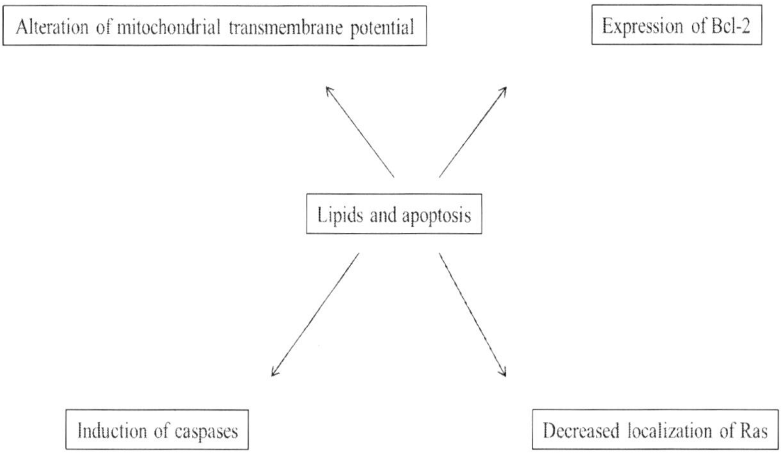

Figure 3.2: The role of lipids in apoptosis.

3.3.3 Lipids and Infections

While the levels of molecules involved in inflammation were lowered by the administration of *n*-3 or *n*-9 fatty acids; a challenge here is the lowering of resistance to pathogens.

- The clearance of *Listeria monocytogenes* was lowered from the spleen or liver when animal models were fed fish oil. Animals fed with lard oil, hydrogenated coconut oil, or palm oil were challenged with a lethal dose of a virulent *L. monocytogenes* that showed an increased liver/spleen clearance as well as increase in cell death due to bacterial activity that could be traced back to a possible dip in IL-12 or inhibition of Ia or MHC Class II molecules in the mice.
- The resistance of mice to *L. monocytogenes* was not altered by conjugated linoleic acid.
- While the clearance of influenza virus was lowered in animals administered a fish oil diet, no effect was detected on the activity of NK cells.

The points to be noted before reaching conclusions is the duration and concentration of fatty acids as well as the cell populations studied.

3.3.4 Lipids and Cancer

- The risk of breast cancer has been lowered by a possible increased consumption of long-chain *n*-3 fatty acids or *n*-9 fatty acids.

- The risk of colorectal cancer was lowered in rats by *n*-3 polyunsaturated fatty acids.
- However, colon or breast tumors are probably associated with increased *n*-6 fatty acids or saturated fatty acids.
- Alaskan and Greenland Eskimos who consume more of *n*-3 polyunsaturated fatty acids are protected more from colon cancer. Studies in animals have pointed out to increased apoptosis by fish oil diets.
- The formation of lesions similar to tumors was shown by the administration of saturated, monounsaturated, and polyunsaturated fat mixture used in Western diets. Animal models revealed changes in activities of cyclooxygenase activity.
- Transformation of mouse epidermal cells was inhibited by the administration of *n*-3 fatty acids (Reviewed by de Pablo et al, 2002).

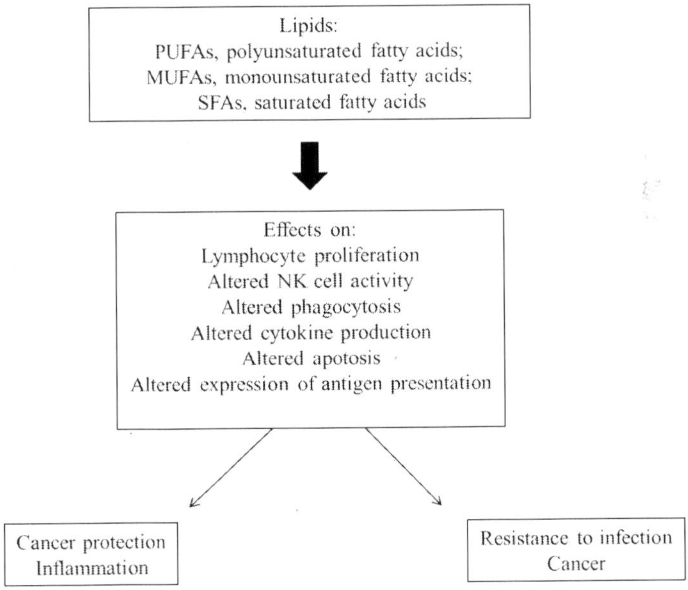

Figure 3.3: Summary of the role of lipids in immune functions.

3.4 BLOOD LIPIDS AND DIETARY FATS

The levels of LDL-C and cholesterol are lowered by omega-6 polyunsaturated fatty acid (PUFA) in place of saturated fat. However, the total lipoprotein profile is not viewed at when this lowered level of LDL is considered. For instance, the impact on cardiovascular disease is more for small-dense LDL (sdLDL). Lauric acid, a saturated fat is associated with an increase in high-density lipoprotein cholesterol (HDL-C) along with total and LDL-C. This calls for challenges in the interpretation of health effects associated with saturated fat or omega-6 PUFA replacements or hence, the dietary fats and fatty acids and their effects on health.

The effect of Monounsaturated fat (MUFA) along with saturated fat in high amounts in a meal increases the levels of chylomicrons and their remnants, known as postprandial lipaemia associated with harm to the arteries. A meal with more saturated fat takes a longer time of the rise in triglycerides or TGs as compared to a faster clearance when the diet contains MUFAs; this serves as an explanation of the benefits offered by MUFA over saturated fat. The oxidation rates of fats are more when a meal is rich in MUFA as compared to saturated fat as well as a lower profile of factor VIIc and factor VIIa (clotting factors) that suggests a lesser increase in weight and thrombosis risk. Trials have revealed a lower profile of factor VII when healthy participants consumed oleic acid-rich meals. The levels of LDL, P-selectin and apolipoprotein B were lower in a Mediterranean Diet; rich in MUFA (DiNicolantonio et al, 2018).

The benefits of MUFAs against fat diets are compared in **Figure 3.4** below:

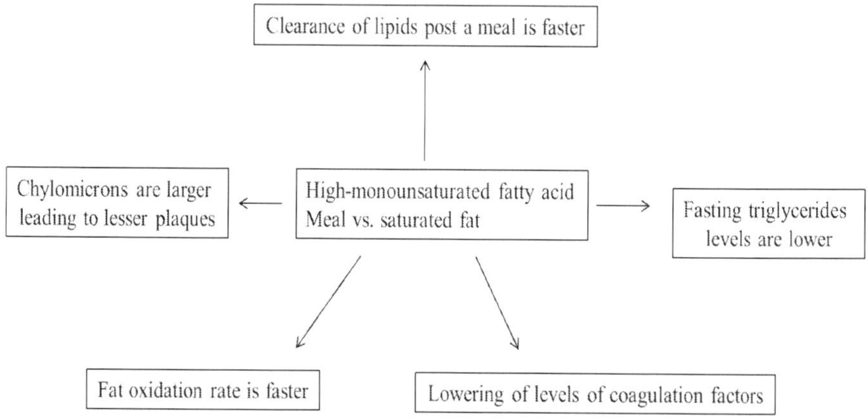

Figure 3.4

A decrease in fasting and postprandial TGs, as well as sdLDL, was seen when a diet consisting of eicosapentaenoic acid (EPA) and docosahexaenoic acid (DHA) at 3:1 was consumed. In a multicenter, randomized controlled study from Europe called LIPGENE isocaloric diets were tested. The diets tested included:

- Saturated fat-rich diet
- MUFA-enriched diet
- High-complex carbohydrate diet
- The above diet with omega-3 PUFAs

The level of LDL was lowered in these groups; the decrease was lowered by 15% in high-fat MUFA rich diet and 37% in the long-chain omega-3 PUFA group; these groups also showed a decrease in TGs (Hartwich et al, 2009). The consumption of marine omega-3 acids is attributed to the benefits of DHA against EPA. The supplementation of EPA also is reflected by lowered sdLDL and C-reactive protein (Reviewed by DiNicolantonio et al, 2018).

This is summarized by Figure 3.5:

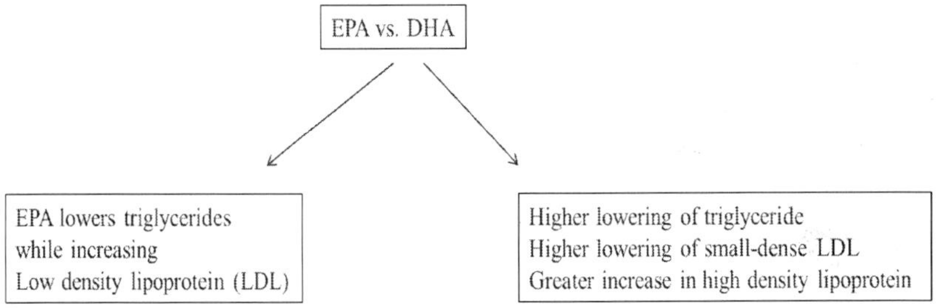

Figure 3.5

3.5 LIPIDS AND SKIN

The metabolism of lipids plays an important role in skin health; with the skin protecting the body from water, solar radiation as well as harmful organisms. The lipid metabolism is distinct on the skin with a varying degree of bioactive lipids. Of the 3 layers on the skin,

- On the skin dermis, corneocytes are present rich in proteins in a matrix of lipids mainly ceramide species, cholesterol with free fatty acids.
- Hypodermis contains adipose tissues containing fatty acids that are mediators.
- The sebum present on the surface of the skin contains

diacylglycerols, wax esters, triacylglycerols, squalene, and esters protects from the sun and microbes.

- The function of linoleic acid (LA; C18:2n-6) for ceramides that function in the epidermal barrier.
- Several PUFAs are involved in inflammation such as docosanoids, eicosapentaenoic acid (EPA; C20:5n-3), endocannabinoids.
- A specialized mix of lipids maintains the functions a barrier. 10% of the skin barrier is made on the basis of lipids whose composition of 50% ceramides, 25% cholesterol and 15% non-esterified fatty acids (Kendall et al, 2017).

The various roles of lipids in the skin are summarized in Figure 3.6

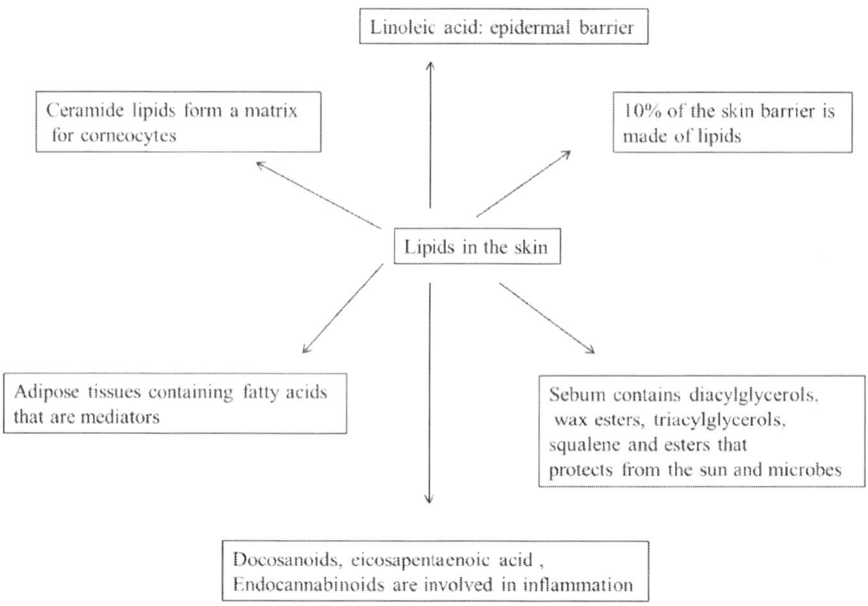

Figure 3.6

The complexity and importance of ceramides have been reported; membrane sheets are formed by these ceramides. In the case of deficiency of fatty acids, the functions of the barrier are compromised. The amelioration of inflammation has been reported by the use of n-3PUFA supplementation. The role of PUFA-derived bioactive lipids was explored by scientists. An article published in 2017 explored the role of supplementing n-3 PUFA on the ceramides of the skin. This was done as ceramides are important for the skin functioning and hence, the effect of supplementation with

N-3PUFA was assessed. The profile of 321 ceramides was assessed by semi-quantitative UPLC/ESI-MS/MS method. Few examples of the assessed molecules include: CER[NS], CER[NH], CER[EOS], CER[AS], CER[NP], CER[NDS], CER[EOH], , CER[AH] and CER[EOP] that are vital for signaling and the formation of the epidermal barrier.

The results are summarized below:

- EPA increased the level of CER[NS] and CER[NDS] in the dermis. The increase due to DHA was not significant statistically for the ceramides analyzed.
- EPA or DHA did not increase other members of the ceramides.
- The n-3PUFA supplementation causes an increase in non-hydroxy fatty acids as there was no difference in the levels of CER[AS], CER[ADS], CER[AH] or CER[AP].
- There was no change in CER[EOS], CER[EOH] and CER[EOP] in the epidermis.
- The supplementation also showed no change of sphingoid bases or phosphorylated sphingoid bases.
- Compartmentalization of functions is suggested as the dermis and epidermis show different responses; EPA increased the level of CER[NS] and CER[NDS] ceramides only in the dermis while C1P was higher in the epidermis.

Thus, the authors reported the effects of n-3PUFA in skin health with implications for further research in the future to study the role of lipids on the skin (Reviewed by Kendall et al, 2017).

3.6 TRANS FATTY ACIDS

The presence of double bonds in the trans configuration earns these fats this name. Monounsaturated, polyunsaturated and saturated fatty acids together make up fats and oils. The presence of a double bond in trans makes Trans fatty acids (TFAs). These are formed during the production of fat by industrial partial hydrogenation of oils. The TFAs in the diet are:

- Vaccenic acid (18:1, trans-11) from ruminants and
- Elaidic acid (18:1, trans-9) obtained from industrial hydrogenation

The various types of fatty acids are summarized in **Figure 3.7**:

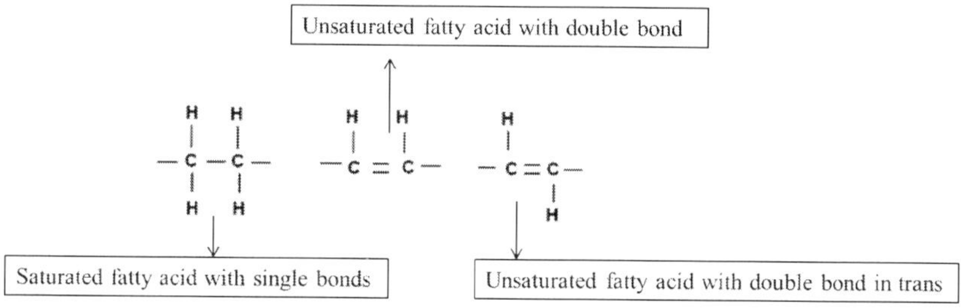

Figure 3.7

Nearly 80% of trans fats in the diet arise from processed foods while natural foods contain 20%. The sources are cookies, animal products, cakes, chips, margarine, shortenings and popcorn. While a lesser than 20% intake of fats and oils is associated with unfavorable HDL and triglycerides profiles as well as Vitamin E levels, there are several risks associated with increased consumption:

- Diets with higher saturated fatty acids are associated with increased serum cholesterol that ups the risks of atherosclerosis or coronary heart disease.
- TFA from industrial sources is linked to higher risk of coronary heart disease.
- TFA from industrial sources lowers levels of LCPUFAs: long-chain polyunsaturated fatty acids important for vision and development of the central nervous system in a fetus.
- The level of trans fatty acid levels in red blood cells was more in women with preeclampsia.
- A 14-year Nurses' Health Study showed an increased trans fatty acid intake and development of type-II diabetes.
- The risk of obesity, allergic cold, and allergy was higher with the consumption of trans fatty acids.

While the use of trans fatty acids is favored in food due to their structures, there are studies that have reported several health risks with their consumption. A banning of trans fats is not a solution as several products from ruminants are beneficial such as iron, calcium, and proteins. Infants too require these products that hence require the adoption of approaches such as:

- Minimizing the intake of trans fats.
- Consumer awareness

- The use of alternative fats in food by hotels and manufacturers.
- Legislation to enforce strict practices by the Governments.

Alternatives to trans fats include:

Modification of the hydrogenation process:

An increase in the degree of hydrogenation lowers the level of TFA with an increase in saturated fatty acids; the process can be modified to allow for functional fats.

Interesterification:

This involves rearrangement of fatty acids within a molecule to keep the composition intact. Thus, this process can avoid the formation of TFA but retain desirable properties. An example is margarine and fat spreads free of trans fats.

Fractions:

Natural oils can be fractionated to yield fractions with high solid content.

Trait-enhanced oils:

An example is low-linolenic oil made of 1–3% α-linolenic acid or 10% α-linolenic acid known as canola oil. The stability to oxidation is increased by using breeding or biotechnology (Dhaka et al, 2011).

The various aspects of trans fats are summarized in **Figure 3.8**:

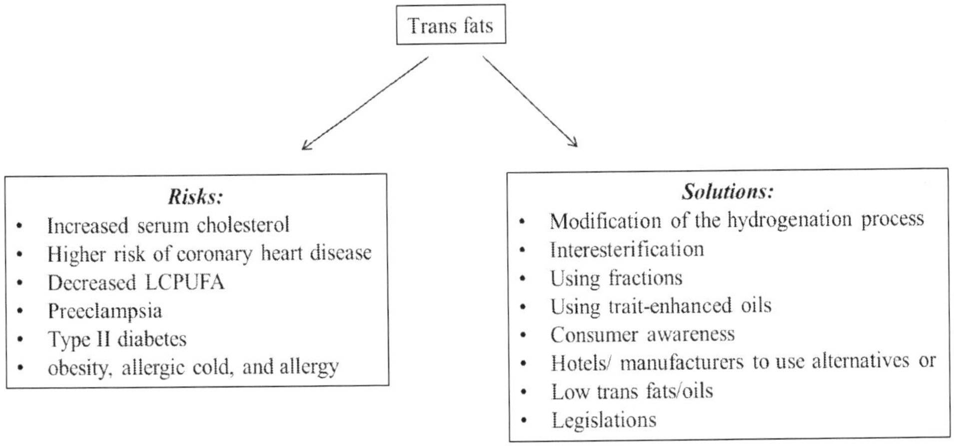

Figure 3.8

3.7 LIPIDS AND THE BRAIN

Polyunsaturated fatty acids (PUFA) of the cell membranes maintain the barrier of hydrophilic compartments. While the hydrophobic portions are oriented towards the bilayers; the hydrophilic ones are exposed to aqueous media. 60% of the dry weight of the brain is constituted by fatty acyl chains. The synthesis of other PUFA requires linoleic acid (LA, C18:2) and alphalinolenic acid (ALA, C18:3) in the diet of mammals. The brain has an abundance of Arachidonic acid (ARA) and docosahexaenoic acid (DHA) along with eicosapentaenoic acid (EPA). As there is an increased tendency of oxidation by double bonds carbon methyl groups, lipid peroxidation or cross-linking would occur in the absence of antioxidants. The recommended intake of EPA+DHA is a minimum of 250 mg. the development of the brain requires ARA and DHA.

The source of LA is wheat, soy, sunflower, and corn while ARA is sourced from meat and eggs. ALA is found in walnut, lingonberry, chia, flax, and rapeseed. EPA+DHA (ω-3 long-chain PUFA) is found highest in fish oil and oily fish. Certain ω-3 fatty acids are obtained by fermenting microalgae.

3.7.1 PUFA and the Human Body

The functions of PUFA are summarized below:

- They function as fuels.
- They are involved in regulating transcription factors.
- They cause an increase in fluidity of membranes and allow functioning of proteins.
- They give rise to eicosanoids and docosanoids that are secondary messengers involved in inflammation and growth, sleep and pain.
- ω-6 fatty acids augment inflammation while ω-3 fatty acids lower it.
- The growth of dendrites depends on long-chain PUFA.
- Neurogenesis in the hippocampus is stimulated by ARA and DHA
- PUFA enhances the plasticity of synapses
- Long-chain PUFAs are involved in the functioning of monoaminergic systems.

A dose of more than 3 g per day of EPA+DHA is associated with inflammatory bowel disease, chronic infections, cancer and disturbances to the nervous system attributed to the altering of lipid rafts and hence functions of signal circuits (Bentsen, 2017)

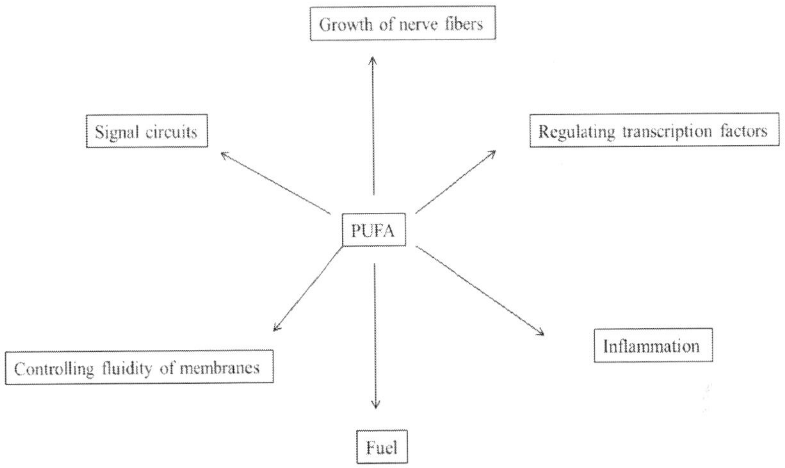

Figure 3.9: Summary of the functions of lipids.

3.7.2 PUFA and Mental Health

- Breast milk contains PUFAs and cognitive development was improved by breastfeeding in a controlled trial (Bentsen, 2017).
- A three-month randomized double-blind trial published in 2014 reported that supplementation of 59 moderately malnourished Mexican children (8-12 years of age) with 270 mg EPA and 180 mg DHA against soybean oil as a placebo increased executive functioning, IQ, visual-perceptive capacity and processing speed. 73% of the children who received the supplements showed large changes against 25% placebos (Portillo-Reyes et al, 2014).
- Another 2014 study showed that brain and cognition showed better results in 65 healthy adults administered EPA 1.3 g day–1 + DHA 0.9 g day–1. The increase in PUFAs increased brain-derived neurotrophic factor, executive functions, lowered decrease in the volume of grey matter and microstructural integrity of the white matter. It was concluded that 'Our findings suggest novel strategies to maintain cognitive functions into old age' (Witte et al, 2014).
- The levels of ω-3 PUFA were lower in people with depression as against controls. The clinical symptoms of depression were

improved (Hedge's G = 0.61, $p < 0.001$) when patients received ω-3 long-chain PUFA against controls in one study (Bentsen, 2017)

- In another study, transition into psychosis was prevented in ultra-high-risk adolescents administered EPA, DHA, and α-tocopherol over an 8-year study (Amminger et al, 2015).
- The level of EPA is lower in minor cognitive deficits while that of ω-3 long-chain PUFA are lesser in dementia. The risk of dementia and decline in cognition is lowered with the consumption of fish and ω-3 PUFA (Bentsen, 2017).
- ADHD or Attention Deficit Hyperactive Disorder is linked to lesser levels of ω-3 fatty acids, folate, and fiber. The levels of DHA are less in ADHD (Bentsen, 2017).
- A study published in 2014 reported that the treatment of young ADHD patients with ω-3 fatty acids resulted in improved attention and improvement in the hyperactivity as reported by teachers and parents (Hawkey et al, 2014).

3.8 FINAL NOTE

In order to highlight the importance of lipids in health, various roles played by lipids have been discussed in this chapter. The role of lipids in healthy hair follicles and skin (especially ceramides) were highlighted. Additionally, the role of various fatty acids in immunity and the role of various lipids on immune functions were discussed. The exact changes in blood lipids when diets containing MUFAs were consumed in order to show the importance of taking the entire lipid profile into account when studying diets.

Additionally, in the current scenario of increased dementia, depression and other mental disorders, the role of PUFA in mental health was also discussed. Overall the effect of lipids in this chapter was not seen as dreadful constituents of the diet rather when taken at prescribed amounts can prevent major illness and keep us healthy.

3.10 REFERENCES

1. Amminger GP, Schafer MR, Schlogelhofer M, et al. Longer-term outcome in the prevention of psychotic disorders by the Vienna omega-3 study. *Nat Commun.* 2015; 6:7934. DOI:10.1038/ncomms8934.
2. Bentsen H. Dietary polyunsaturated fatty acids, brain function and mental health. Microbial Ecol Health Dis. 2017; 28:1281916.
3. de Pablo MA, Puertollano MA, de Cienfuegos GA. Biological and Clinical Significance of Lipids as Modulators of Immune System Functions. *Clin Diag Lab Immunol.* Sept. 2002; 945–950.
4. Dhaka V, Gulia N, Ahlawat KS, Khatkar BS. Trans fats—sources, health risks and alternative approach - A review. *J Food Sci Technol.* (September–October 2011); 48(5):534–541.
5. DiNicolantonio JJ, O'Keefe JH. Effects of dietary fats on blood lipids: a review of direct comparison trials. *Open Heart.* 2018;5:e000871.
6. Evers BM, Farooqi MS, Shelton JM, et al. Hair growth defects in Insig-deficient mice caused by cholesterol precursor accumulation and reversed by simvastatin. *J Invest Dermatol.* 2010; 130:1237– 1248.
7. Hartwich J, Malec MM, Partyka L, et al. The effect of the plasma n-3/n-6 polyunsaturated fatty acid ratio on the dietary LDL phenotype transformation - insights from the LIPGENE study. Clin Nutr 2009;28:510–5.
8. Hawkey E, Nigg JT. Omega-3 fatty acid and ADHD: blood level analysis and meta-analytic extension of supplementation trials. *Clin Psychol Rev.* 2014; 34:496–505.
9. Karnik P, Tekeste Z, McCormick TS, et al. Hair follicle stem cell-specific PPARγ deletion causes scarring alopecia. *J Invest Dermatol.* 2009; 129:1243–1257.
10. Kendall AC, Kiezel-Tsugunova M, Brownbridge LC. Lipid functions in skin: Differential effects of n-3 polyunsaturated fatty acids on cutaneous ceramides, in a human skin organ culture model. *Biochimica et Biophysica Acta.* 2017; 1859(2017):1679–1689.
11. Muro E, Atilla-Gokcumen GE, et al. Lipids in cell biology: how can we understand them better? *Mol Biol Cell.* 2014; 25:1819-1823.
12. Portillo-Reyes V, Perez-Garcia M, Loya-Mendez Y, et al. Clinical significance of neuropsychological improvement after supplementation with omega-3 in 8-12 years old malnourished Mexican children: a randomized, double-blind, placebo and treatment clinical trial. *Res*

Dev Disabil. 2014; 35:861–870.

13. Stenn KS, Karnik P. Lipids to the Top of Hair Biology. *J Invest Dermatol*. 2010 May; 130(5): 1205–1207.
14. Wan Y, Saghatelian A, Chong L-W, et al. Maternal PPARγ protects nursing neonates by suppressing the production of inflammatory milk. *Genes Dev*. 2007; 21:1895–1908.
15. Witte AV, Kerti L, Hermannstädter HM, et al. Longchain omega-3 fatty acids improve brain function and structure in older adults. *Cereb Cortex*. 2014; 24:3059–3068.

CHAPTER 4

LIPIDS IN DISEASE AND TREATMENT

Shiv Sanjeevi Sripathi

Shiv Sanjeevi Sripathi secured second rank in Mumbai University and first rank in his college at his undergrad level and in his college during his Masters and has worked on a stem cell project at the Specialized Centre for Cell Based Therapy (SCCT), KEM Hospital at Mumbai, following which he was a Junior Research Fellow at Microbiology & Cell Biology Department, Indian Institute of Science, Bangalore from August 2009 to November 2011. The field of work involved cloning of cell wall genes and transcription factors in E. coli & M. smegmatis. This was followed by a stint as a Research and Development Executive at Transasia Biomedicals, Mumbai where he worked on patent writing and Immunoassays from 2011 till 2013. In 2013, he entered the teaching field where he has been tutoring biology and chemistry to the students of 11th and 12th grades along with English grammar and communication skills to people of all age groups. Shiv also loves helping people write scientific manuscripts and has co-edited several books on life sciences and aspects of biotechnology with Apple Academic Press.

4.1 LIPIDS: WHAT THEY DO?

Many vital functions have been attributed to lipids that comprise of:

- Function as fuel: storage and transported forms
- They are components of membranes
- Essential fatty acids of the diet are nutrients
- They are associated with components on the surface of cells that are involved in recognition as well as immunity.
- They play important roles in food texture (discussed in Chapter 2).
- They are carriers of Vitamins A, D, E and K (fat soluble vitamins).

80% - 95% of lipids are triacylglycerols (TAG) that are liquid (oil) or solid (fat) at room temperature. Fatty acids are monounsaturated (MUFA) and polyunsaturated (PUFA): the latter is of 2 types n-3 (also omega-3), n-6 (also omega-6) based on the first double bond from the methyl end group of the fatty acid. The contents of adipose tissues and circulatory lipids matched with the dietary lipids of seals and fish in sync with "we are what we eat".

4.1.1 Short-Chain Fatty Acids (SCFA)

The fermentation of carbohydrates in the human gut yields C 2:0 to C 4:0 or Short-chain fatty acids (SCFA). The calories they provide is lower and they dissolve to a greater extent in water allowing their rapid absorption.

4.1.2 Medium-Chain Fatty Acids (MCFA)

They are made of 6-12 Carbon atoms and are hydrolyzed products of plant oils. These acids have low melting point and viscosity as well as stability to oxidation. Several properties are summarized below:

- Due to their high solubility, their absorption into the portal system does not involve hydrolysis and re-esterification.
- The rate of β-oxidation is higher.
- Chylomicron formation is not required for transport.
- Their tendency to accumulate as body fat is very low.
- The calorific value is more than other nutrients while their metabolism and absorption are on the lines of glucose.
- Their entry into mitochondria does not require carnitine.
- The metabolism, transport as well as absorption are faster than

LCFA to allow for absorption.

- A combination of these fatty acids with long-chain triacylglycerols (LCT) is utilized in medical foods. The digestion of LCT is challenging with cystic fibrosis, Crohn' disease, enteritis, etc.
- The level of cholesterol in the serum and tissues is lowered by MCT like polyunsaturated oils according to certain reports.

4.1.3 Long-Chain Fatty Acids

Long-chain triacylglycerols (LCT) contain more than >C12; examples are stearic acid (C18:0) and palmitic acid (16:0). On account of their higher melting points, their absorption is compromised. This causes the design of structured lipids (SL) with a modified backbone. An example is Salatrium: made of stearic acid and SFCA (Finley et al, 1994).

4.1.4 Essential Fatty Acids (EFA)

When PUFA contains more than one double bond; the synthesis is not possible in the body that makes them EFA. There are 2 categories: the n-3 and the n-6 fatty acids with parent compounds linoleic acid (LA, 18:2 n-6) and α-linolenic acid (ALA, 18:3 n-3), respectively.

The n-3 Fatty Acids

Examples include α-linolenic acid (ALA), eicosapentaenoic acid (EPA; 20:5n-3) and docosahexaenoic acid (DHA; 22:6n-3). Several health benefits have been reported such as:

- High intake of n-3 PUFA in Greenland Eskimos and their lower risk of coronary heart disease.
- EPA lowers the level of cholesterol in the serum and yields a series of eicosanoids that bestows against heart attacks.
- The protective role of n-3 fatty acids on heart health and cardiovascular disease (CVD) has been studied.
- DHA is vital for the central nervous system (CNS) and visual acuity in infants.
- The grey matter of the brain contains DHA.

As the susceptibility of long-chain PUFAs to oxidation is higher, the use of appropriate packing and antioxidants is warranted.

The n-6 Fatty Acids

These fatty acids are involved in regulating the transport of cholesterol and maintenance of skin permeability as well as structures of membranes. Linoleic acid (LA; 18:2n-6) is a ubiquitous example that is present in vegetable oils to function as a precursor of arachidonic acid (AA; 20:4n-6) that is a vital phospholipid of the membrane. An intermediate between LA and AA is γ-linolenic acid (GLA; 18:3n-6) that is present in several oils. The consumption of 1% - 2% LA protects against several diseases such as compromised healing of wounds as well as affected reproduction and growth (Hamam, 2013).

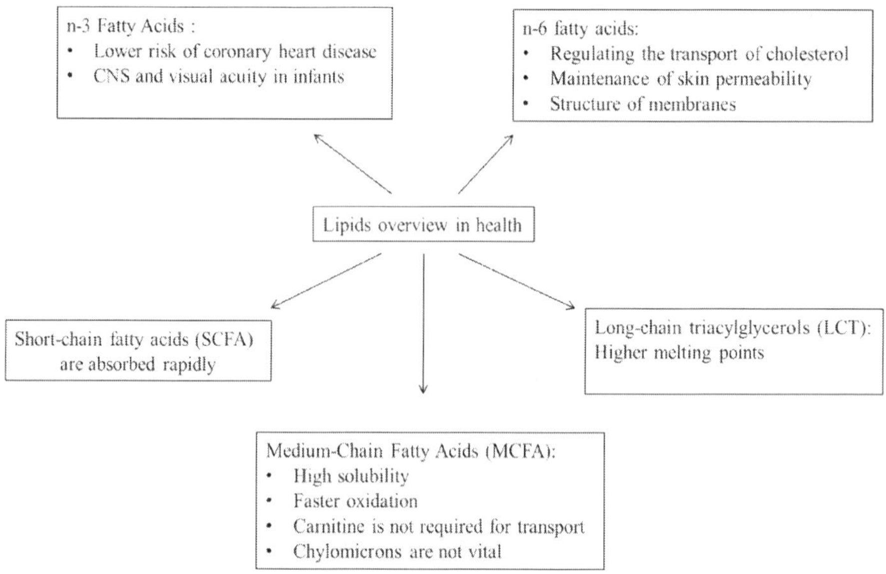

Figure 4.1: A summary of various health benefits of lipids.

4.1.5 Structured Lipids

As previously discussed TAGs are modified in structured lipids (SL) using chemical methods or enzymes. The benefits that look promising include (Hamam, 2013):

- Protecting from thrombosis
- Boosted immune systems
- Lowered TAG and sterol
- Lowered breakdown of fats

- Lowered cancer risk
- Preserving of functions of the reticuloendothelial system
- Lowered calories
- Improved Nitrogen balance

The next few sections discuss the involvement of lipids in various diseases.

4.2 LIPIDS IN SKIN DISEASE: PSORIASIS

Psoriasis is described as an auto-immune disorder involving "erythematous scaly plaques over extensor aspects of the body". 2-3% of the population is a target of this disease. The risk of cardiovascular events is higher in long term sufferers that are attributed to lowered anti-oxidant activity, higher oxidative stress as well as obesity and hypertension. The role of lipids in this disease was assessed in this work published in 2016 where the lipid profile was compared in 30 healthy and 30 individuals with the disease.

The authors observed that the levels of triglyceride (TG), total cholesterol, low-density lipoprotein, and VLDL in the serum of patients with severe disease than the moderate disease. The level of HDL–C had a significantly higher value in mild cases than controls. The total cholesterol/ HDL ratio is more in the patients than controls.

While the exact reason behind the changes in lipid profile in the case of psoriasis lacks clarity as to whether the change is due to the disease as such or associated events. The association between psoriasis and dyslipidemia is yet conflicting with some studies showing more than 50% patients having dyslipidemia while others did not.

The observed increase in lipids in psoriasis is explained by several hypotheses; the disease has been classified as a systemic inflammatory disease with the involvement of cytokines. The role of platelet adhesion by VLDL in atherosclerosis has been pointed out in literature. Additionally, the swallowing of LDL by macrophages stimulates the production of tumor necrosis

Factor (TNF-α) and IL-1β that are cytokines associated with inflammation. The association of lipid changes in immunology in psoriasis is associated with an "immunometabolic" syndrome like psoriasis. Overall the risk of cardiovascular diseases has been reported as higher in psoriasis. Thus, changes in lipids in serum can serve as prognostic markers to assess the disease-associated risks (Gupta et al, 2016).

4.3 CARDIOVASCULAR DISEASE RISK AND LIPIDS: A CASE STUDY

The search for new markers to assess and predict future cardiovascular events is on due to the current set of markers that have not augmented the prediction of risks. The development of cardiovascular disease (CVD) is connected with plasma triglycerides, cholesterol with LDL- and HDL-cholesterol used as benchmarks. However, as these measurements are gross and do not entail the study of specific lipids involved in plaque formation a hypothesis was tested as to alterations in particular lipid species associated with predicting CVD risk than the conventional total measurements. This involves the use of the emergent field of lipidomics where the entire set of lipids in a biological sample is described at a particular time. This involves the use of a shotgun technique where a crude lipid extract is subjected to mass spectrometry (MS) analysis. This technique is reported as reproducible, sensitive, automated and with a high-throughput. Individual lipid species were identified along with 23 gene variants for coronary disease risk.

As the potential of the lipidome remains unexplored, the authors used top-down shotgun lipidomics in a prospective population-based cohort to assay for variations in the composition of the plasma in patients with similar lipid profiles. This study was reported as a first to assess plasma profiling for lipids for CVD prevention; 427 samples were analyzed (2 replicates) to identify 85 lipid species belonging to 9 different classes of lipids.

As TAG levels have been used as risk factors, two specific LPC species i.e., LPC 16:0 and LPC 20:4 among the lysophosphatidylcholine (LPC) family were shown to be protective in CVD. The authors thus, suggest analyzing lipid species over whole profiles to gain deeper insights into lipids in CVD. Other results are:

- Carotid intima-media thickness showed a negative relation with LPC16:0.
- Increased association of CVD was revealed by Sphingomyelin (SM) 38:2.
- LDL-cholesterol and SM 38:2 showed a positive correlation.
- Protection was seen in five triglyceride (TAG) species.
- Among the gene variants linked to circulating lipid species concentrations, LPA, SORT1 and ZNF259/APOA5-A4-C3-A1 have been associated with lipoproteins while 3 are reported for the first time here: WDR12, ZC3HC1, and PHACTR1.

- LPC16:0 and/or LPC 20:4 were lower in risk allele carriers of 3 CAD-loci.

The authors described the study as "a proof-of-concept screen that shotgun lipidomics can be used as a tool in the search for novel CVD biomarkers". The use of individual lipid species over whole analysis was also suggested as well as an association of genetic susceptibility. As there was a loss of statistical significance, the authors recommend larger studies (Fernandez et al, 2013)

4.4 LIPIDS AND KIDNEY DISEASES

Chronic kidney disease (CKD) patients show high mortality due to cardiovascular disease (CVD) that entails the identification of new factors to determine and manage the CVD deaths in these patients. CKD patients manifest oxidative stress, uremic toxins, hyperlipidemia as well as inflammation that potentially augment the formation of atherosclerotic plaques. Studies have pointed out at a declining kidney function and higher levels of low-density lipoprotein (LDL) cholesterol with malfunctioning high-density lipoprotein (HDL) cholesterol.

The transport of insoluble triglycerides and cholesterol molecules is accomplished by a coat of lipoproteins or amphipathic lipids and apo-lipoproteins. These lipoproteins are of several types: HDL contains maximum protein while chylomicrons contain excess lipids. Very low-density lipoprotein (VLDL) particles transport lipid from the gut to the tissues and muscle while HDL mediates elimination via the liver.

Alterations in lipoproteins are associated with renal dysfunction:

CKD shows a deficiency of lecithin-cholesterol acyl-transferase (LCAT) that prevents the maturation of HDL to manifest as:

*Low HDL levels

*Higher LDL

*Increased levels of lipoprotein particles

Overall this contributes to oxidative stress and higher mortality due to cardiovascular events in CKD patients

Nephrotic Syndrome shows an alteration in the metabolism of lipids and lipoproteins. There are changes as below:

*The level of HDL is similar to healthy individuals

* VLDL and intermediate-density lipoprotein (IDL) levels are higher

The antioxidant and anti-inflammatory properties of HDL cholesterol prevent atherosclerosis by lowering the infiltration of monocytes. A deficiency of LCAT leads to compromised maturation of HDL in CKD patients. Additionally, upregulated acyl-CoA cholesterol acyltransferase renders HDL ineffective from lowering the levels of LDL-cholesterol. The reverse cholesterol transport by HDL (transport to the liver of peripheral cholesterol) prevents uptake of peripheral cholesterol by macrophages to form foamy cells. In the case of ApoA-1 deficiency, cholesterol is taken up to form foamy cells that form atherosclerotic plaques. The reactive oxygen species (ROS), as well as infiltration of monocytes, is lowered by HDL cholesterol. Hence, the loss of function or altered function of HDL cholesterol in CKD patients causes an increase in the risk of atherosclerosis.

The levels of VLDL cholesterol, IDL cholesterol, and triglycerides are higher in patients with nephrotic syndrome. Deficiency of hepatic lipase that converts IDL to LDL cholesterol in these patients causes an increase in atherogenic IDL cholesterol and higher triglyceride content in the LDL cholesterol. The formation of small dense LDL (sdLDL) in CKD patients is linked to improper HDL cholesterol along with deficiency in LDL receptor-related protein (LRP). Hence, despite normal levels of LDL cholesterol, atherosclerotic plaques can be formed by IDL cholesterol and sdLDL cholesterol. This dictates a detailed analysis of lipids in CKD patients.

Other lipids that are altered in kidney diseases are serum triglycerides with an increase in triglyceride levels in kidney diseases (the highest seen in nephrotic syndrome). This is due to lipoprotein lipase (LPL) inactivation that leads to lowered triglyceride catabolism.

A challenge is low or normal levels of serum cholesterol levels in CKD patient. Despite the low level, the mortality is highly linked to inflammation and oxidative stress and increased oxidized LDL cholesterol. Hence, a thorough analysis of the lipid profile is warranted in the case of CKD patients. While hemodialysis patients show higher predicted morbidity and mortality due to low cholesterol with inflammation, a linear relation is seen in the case of in non-dialysis CKD patients.

4.4.1 Lipidomics and CKD

The diagnosis, as well as prognosis of kidney diseases, can be facilitated by lipidomics. Examples of such studies that have outlined few biomarkers include:

- Higher amounts of free fatty acid, glycerolipid, and

glycerophospholipid in advanced CKD patients than controls,

- Higher methyl hexadecanoic acid and 3-oxooctadecanoic acid levels linked with lower eGFR levels.
- HigherLPC(24:1)andlowerPC(20:2/24:1)glycerophospholipids.
- Higher palmitic acid and monounsaturated acid levels
- Lower levels of polyunsaturated acid

4.4.2 Treatment Strategies

- Inhibiting Proprotein convertase subtilisin/kexin type 9 (PCSK9) lowers the risk of CVD; the liver enzyme lowers the amount of LDL cholesterol receptor to lower its uptake and hence, increase the levels of LDL cholesterol in the blood.
- The levels of LDL cholesterol are lowered by the use of statins while there has been no change in mortality.
- The levels of triglyceride can be lowered by fibrates in nephrotic syndrome, though the side effects need further exploration.
- The progression of CKD can be slowed by the use of antioxidants to target inflammation and oxidative stress. For example, randomized, placebo-controlled trials have shown lowered CVD in hemodialysis patients by Vitamin E (though larger sample size is warranted).

Thus, the role of several lipids: decreased HDL cholesterol as well as increased SdLDL cholesterol or oxidized LDL cholesterol levels have been highlighted in kidney diseases. The potential of lipidomics to uncover new biomarkers has also been suggested.

The link between atherosclerosis and lipid profiles is summarized in Figure 4.2:

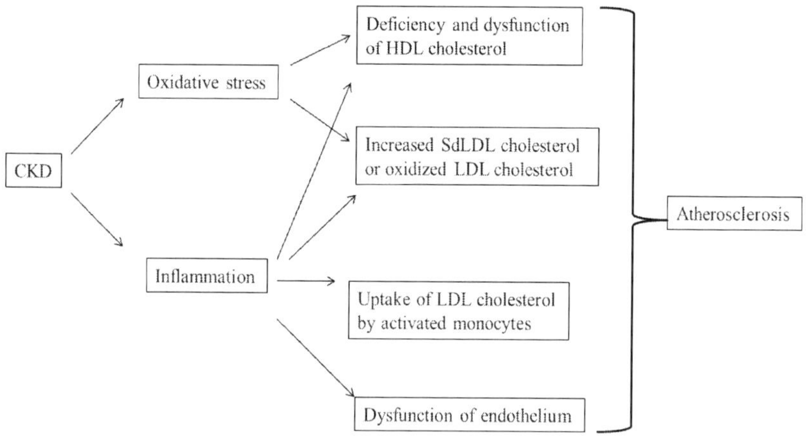

(Bulbul et al, 2018)

Figure 4.2

4.5 DENGUE FEVER AND LIPIDS

Dengue virus (DENV: 4 serotypes; DENV-1, DENV-2, DENV-3, and DENV-4) is transmitted by *Aedes* mosquitoes, for example, *Aedes aegypti* and *A. albopictus*. This is classified as "the most hazardous arbovirus in the world" due to estimates of 390 million infected persons annually (WHO). While the manifestations can range from infection to fatality, several symptoms include fever with chills, tiredness, retro-orbital headache and thrombocytopenia among others. Subsequent infection by a second serotype can lead to severe dengue leading to organ failure in few patients. The development of novel tests for diagnosis or vaccines is limited by paucity in markers. The importance of lipids in the infection process promoted a research team to examine the lipidome of DENV-4 infected patients with minimum preparation using direct infusion high-resolution mass spectrometry instead of liquid chromatography coupled to mass spectrometry or LC-MS.

The infection of DENV is associated with changes in lipids when the replication of the virus involves control of autophagy of the host cell. The demands for viral replication are met by the virus by controlling autophagy and hence, metabolism of lipids. A reorganization of the membranes of host allows for the production of lipid vesicles to form viral replication complexes. The viral replication is associated with an increase in phosphatidylcholines (PC). The lipidome analysis revealed 3 PCs associated with viral replication seen as distinct features in patient sera. The data were reported as in agreement of positive-chain RNA viruses as an *in vivo* proof. In order to meet the increased demand for energy, the virus controls Triglycerides or TGs and cholesteryl esters present in lipid droplets. The study identified TGs: TG(44:4), TG(46:5), TG(48:6) and TG(52:2) as markers for infected patients.

The study also reported 4 PAF (Platelet activation factor) precursors for use as biomarkers; PAF is associated with immune and inflammatory cascades in platelets as well as macrophages, eosinophils and neutrophils. DENV infection is associated with the release of PAF causing the inflammatory conditions. The manifestation of thrombocytopenia is seen as a classical measure of the disease caused by PAF- DENV interaction.

Overall, the authors reported changes in lipid profiles in response to DENV seen as changes in PC and TGs. The increase in PAF was in line with manifestations associated with this molecule thrombocytopenia and hemorrhagic fever (Melo et al, 2018).

4.6 SERUM LIPIDS AND MALARIA

Tropical countries are challenged with the occurrence of malaria due to *Plasmodium vivax* (85% cases). Several diseases of an inflammatory or infectious nature are associated with alterations in the profile of serum lipids. This is due to lipids being associated with the host and *Plasmodium* with metabolic requirements as cholesterol and phospholipids as well as modified pathways in the host. For example, malaria is associated with hypocholesterolemia, higher levels of triglycerides, and very low-density lipoproteins (VLDL) and decrease in high-density lipoproteins (HDL) and low-density lipoproteins (LDL). While the majority of the studies have been performed on adult populations, the changes in children have not been investigated exhaustively. A research article published in 2016 explored the profile of lipids in sera of children (43 in number) with low-level *P. vivax* infection vs. healthy controls (32 in number).

The results found are given below:

- The levels of cholesterol were normal, but the levels were lesser in the patients than controls. The decrease in cholesterol in malarial patients can be associated with metabolic requirements of *Plasmodium*.
- The level of LDL and HDL in patients at the time of admission was lower than the controls. A similar profile was seen in falciparum

malaria affected Nigerian children. Such changes in lipids are reflective of such inflammation.

- The children with malaria showed an increase in triglycerides and VLDLs. This is attributed to a lowering in the clearance of the lipoprotein.

The lipid levels were also associated with changes in altered lipid profiles as well as ending of the symptoms and fever. The alterations in the lipid profile in children with low levels of infection. These alterations were described as transient with no connection with the patient history (Dias et al, 2016).

4.7 LYSOSOMAL STORAGE DISEASES (LSDS) AND LIPIDS

These are genetic and metabolic disorders attributed to defects in acidic hydrolase or lysosomal membrane proteins that cause an accumulation of undegraded products. Most (70%) LSDs are associated with the central nervous system (CNS). The involvement of lipids is summarized below:

- The stability and curvature of the sheath of neurons are facilitated by galactolipids.
- Compromised or impaired synthesis or transport of galactolipids is associated with progressive motor dysfunction such as amyotrophic lateral sclerosis (ALS), Menkes disease and Pelizaeus–Merzbacher disease (PMD).
- The 200 odd molecules of ceramides differ in structure in mammals and serve as the basis of Glucosylceramide, sphingolipids, and gangliosides. Accumulation of these ceramide lipids leads to apoptosis.
- The levels of ceramides are higher in sporadic ALS (sALS) patients as well as mice models of ALS.
- A rise in ceramides is seen in ataxia telangiectasia (AT) and hereditary sensory neuropathy type 1 (HSN1).
- The development and functioning of the brain require Glucosylceramide (GlcCer); Gaucher disease (GD) is the manifestation when glucocerebrosidase (GBA) of the lysosome is mutated.
- The levels of Glucosylceramide are higher in spinal cords of the

paralytic amyotrophic lateral sclerosis (ALS) patients.

- The level of GlcCer is lower in the peripheral nervous system of Charcot–Marie–Tooth disease Type III.
- Mutations in the enzyme that breaks down GalCer and galactosylsphingosine called galactocerebrosidase leads to Krabbe disease (KD) or globoid cell leukodystrophy.
- Sphingomyelin phosphodiesterase 1 (SMPD1; hydrolyzes its substrate sphingomyelin that is a component of membranes among other cellular functions to ceramide) deficiency is seen in Niemann-Pick disease.

Overall, as reported there are several metabolic aberrations due to mutations in a lysosomal enzyme. The progress of diseases can be facilitated by the alterations in the various lipid classes. The identification of altered lipid profiles or composition or properties of a targeted lipid can serve to see which part of the body would be affected and also to look at novel treatment approaches (Reviewed by Dodge, 2017).

4.8 CANCER LIPID METABOLISM

The malignant properties of cancer have been associated with alterations in metabolism. Alterations in lipid metabolism in cancer cells have received focus; these cells alter the metabolism of lipids in order to suit their increased requirements. 90% of cancer mortalities are due to metastasis that involves several metabolic changes including that of lipids. Several alterations in lipid metabolism are summarized below:

4.8.1 Anabolism

- ATP citrate lyase (ACLY) converts citrate to oxaloacetate and cytosolic acetyl-CoA. Higher amounts of this enzyme were reported in gastric adenocarcinoma tissues against healthy controls. It also interacted with low molecular weight isoform of cyclin E (LMW-E) implied in migration and invasion of breast cancer. This opens an avenue for addressing cancers with high levels of this cyclin by targeting ACLY.
- The aggressiveness of hepatocellular carcinoma (HCC) was associated with higher levels of Acetyl-CoA carboxylase. As it was linked with disease recurrence and poor survival, this enzyme could serve as a biomarker.

- The progression of malignancy has been linked with Fatty-acid synthase in several studies involving ovarian cancer, HCC, colorectal and prostate cancers.
- The expression of Stearoyl-CoA desaturase (SCD isoform 1) was associated with the stage and metastasis of renal cell carcinoma.
- An overexpression of Diacylglycerol acyltransferase lowered the levels of phospholipids while increasing TGs in human lung SV40-transformed fibroblasts and lowered invasive abilities of the neoplastic cells.

4.8.2 Catabolism

The target of the rate limiting enzyme in catabolism: Monoacylglycerol lipase (MAGL) using shRNA or inhibitors affected the aggressiveness of prostate cancer. The inhibition of colorectal cancer was seen with the use of the same inhibitor.

The functions of Phospholipase A2 were higher in epithelial ovarian cancer.

Metastasis of lymphoma cell was augmented by Phospholipase D (PLD) inhibiting this enzyme resulted in lower adhesion, migration, neovascularization and growth of a tumor.

4.8.3 Lipid Rafts

These are lipid domains made of sphingolipids and cholesterol in cell membranes that function in adhesion, survival, metabolism of lipids as well as metastasis. Cancer has been shown to be influenced by changes in the membrane cholesterol. An example of a receptor involved in invasion, migration, and metastasis is CD44. The spread of breast cancer was controlled by palmitoylation of CD44. The receptor was shed off in pancreatic cancer and gliomas by methyl-β-cyclodextrin (MβCD) that targets lipid rafts. Metastasis in lung cancer was supported by overexpressing an important enzyme in synthesis of cholesterol called squalene synthase. Thus, enzymes in lipid or cholesterol synthesis have roles in cancer aggression.

4.8.4 Metabolism

Optimal elevation of ROS allows cells to proliferate and differentiate while in excess oxidative damage and death follow suit. The switch between catabolism and anabolism between cancer cells and host cells can allow for oxidative phosphorylation to fuel metastasis. A co-culture of adipocytes with ovarian cancer cells allowed the transfer of lipids from the former to the latter.

Progression of malignancy is facilitated by lipid metabolism that generates signals and intermediates. Inhibiting key enzymes of lipid metabolism can allow suppression of metastasis. A few examples of such targeting are shown in Table 4.1:

Metabolism	**Enzyme**	**Inhibitor**
Anabolism	ACLY	SB-204990, LY294002
	ACC	AICAR, Metformin
	FASN	Orlistat: approved by FDA
Catabolism	PLA2	Ecopladib, Giripladib, Darapladib, Varesp-ladib methyl, FKGK11
	MAGL	JZL184, JJKK048
	PLD	VU0359595, FIPI, VU0364739

Key: ACLY ATP citrate lyase, ACC acetyl-CoA carboxylase, FASN fatty-acid synthase, MAGL monoacylglycerol lipase, PLA2, Phospholipase A2, PLD Phospholipase D (Reviewed by Lou et al, 2017).

4.9 CANCER AND LIPID DROPLETS

Lipid droplets (LDs) referred to as the storage compartments of fats in metabolism have been the focus of attention. The spectrum of their functions has been extended from storage to metabolism of lipids, synthesis of membranes, signaling, metabolism as well as cancer. The excess of fatty acids that are not esterified and cholesterol are stored in these organelles to avoid toxic effects on the cells. These spherical organelles are of the range of nanometers to micrometers and are composed of cholesteryl esters, triglycerides containing saturated or unsaturated fatty acids. They are surrounded by a layer of phospholipids (phosphatidylcholine) with proteins. The composition and amounts are reflective of fasting or rest or status. They are present in the cytoplasm as well as nuclei commonly with mitochondria and endoplasmic reticulum.

Initial work showed a richness of lipids in carcinomas that were not characterized. Apart from increased glycolysis, cancer cells also increase synthesis of lipids via fatty acid synthase and increased uptake for oxidation by mechanisms that yet lack clarity. Cancer is associated with increased synthesis and catabolism of lipids from their storage centers. Proteomics revealed higher levels of lipolytic enzymes in aggressive cancers as compared to non-aggressive ones. The formation of LDs allows for storage of the excess amounts of lipids thus released in the cytosol. The amount of LDs in cancer cells is more than that of healthy cells for example in human colon adenocarcinoma. Higher amounts of LDs are seen in breast cancer associated with estrogen/progesterone receptors which in turn are linked to lipogenesis and metastasis. Breast carcinomas rich in lipids commonly lack estrogen receptors that are suggestive of a not so simple mechanism of lipid storage.

Increased amounts of LDs are linked to increased aggressive behavior and resistance to chemotherapy. Cancer cell LDs have been shown to contain caveolins, extracellular signal-regulated kinase 1 and 2 and phosphatidylinositol-4,5-bisphosphate 3-kinase (PI3K) associated with tumor features. The presence of molecules associated with inflammation in LD has been reported. The presence of COX-2 and PGE2 (a prostaglandin seen in several malignancies) has been reported in LDs. Research has shown epidermal growth factor receptor or EGFR activating PLIN protein associated with LD in colorectal cancer.

According to the cancer stem cell hypothesis, there are cancer stem cells (CSCs) in a tumor that allow for self-renewal and generate cell populations. These CSCs show metabolic alterations and in this context, it refers to upregulated lipid synthesis. The number of LDs in immortalized rat cells with overexpressed S18-2: a mitochondrial ribosomal protein was high. Ovarian CSCs have shown higher amounts of LDs containing unsaturated lipids. The LDs has higher levels of stearoyl-CoA desaturase-1 that suggests an altered lipid profile of these CSCs. LDs were higher in colorectal CSCs as well as circulating tumor cells (CTCs) in the blood of patients with metastatic lung or prostate cancer.

Thus, the role of LDs has been extended from storage to cancer that can open up new dimensions of tumor research and the possible therapeutic intervention (Tirinato et al, 2017).

4.10 LIPIDS FOR HEALTH

Food is termed functional when it allows for augmenting good health and lowering the risks of disease. In this context, functional lipids can be described as similar to components of a regular diet but apart from their basic function in nutrition confer physiological benefits or/and lower the risks of disease. Functional lipids include the following:

- Omega-3 cannot be synthesized in the body: while docosahexaenoic acid; DHA is present at high levels in the brain, eicosapentaenoic acid; EPA gives rise to prostaglandins involved in inflammation and clotting. Omega-6 fatty acids: Gamma linoleic acid is obtained from metabolism of linoleic acid that is converted to arachidonic acid (20:4 (x-6); AA) that is linked with phospholipids of the membrane and eicosanoid signaling compounds.
- Conjugated linoleic acid or CLA are linoleic acid isomers, for example, cis-9, trans-11 CLA in ruminant milk and meat. The use of CLA isomers has shown promise to prevent cancer in animal studies.
- Medium chain triglycerides or MCTs contain fatty acids having six to twelve carbon atoms. They are digested with ease in the absence of bile salts while their oxidation does not require carnitine. They thus, serve as a source of lipids for patients with abnormalities in fat processing and absorption.
- Phytosterols (plant sterols) are biologically active and possess a structure similar to cholesterol. They impair absorption of cholesterol to hence control its levels and hence associated risks. Examples are stigmasterol, campesterol as well as beta-sitosterol.

The role of lipids in preventing or alleviating symptoms of various diseases is summarized below:

- Depression ranks fourth as the burden of disease in the world with 121 million people affected globally, according to WHO. A significant portion of the brain is composed of lipids: DHA, AA, sphingolipids, cholesterol to 36-60% of its weight. As discussed in Chapter 3, DHA has important roles in regulating gene expression, signal transduction, growth of neuritis as well as events associated with membranes. Studies point at the lowered incidence of depression and PUFA. In a trial published, the administration of 4g EPA over 9 months lowered sadness, low self-esteem, insomnia, appetite and tension (Puri, 2005).
- Discoloration, itchiness, flaking, and bleeding of skin is seen in Atopic dermatitis (AD). The importance of EFAs in the physiology of skin has been documented. The administration of

CLA in pomegranate seed oil lowered allowed strengthening of the epidermis.

- The metabolism and health of bones are facilitated by long chain PUFA and omega-3 fatty acids.

The quality of life, reduced disease risk and good health is allowed for by functional lipids. While the benefits conferred by functional lipids have been reported, further clinical trials are warranted to assess potential benefits and side effects (Alabdulkarim et al, 2012).

4.11 LIPIDS IN THE BRAIN

The energy requirement of the brain is very high with a majority of the requirement met by glucose and fatty acids accounting for $1/5^{th}$ of this requirement. While oxidation of fatty acids increases during fasting, a challenge is the generation of superoxides that damage the organ. An alternative to this pathway is to form ketone bodies from acetyl-CoA. The lipids of the brain play a vital role in structure of cell membranes. Examples are sterols, phospholipids and sphingolipids; the phospholipid bilayer of membranes is vital.

- Among these phospholipids, the major component is phosphatidylcholine with choline serving as a neurotransmitter. The fluidity of the membrane is increased with the levels of phosphatidylethanolamine that also allow for fusion of membranes and vesicular budding.
- The phosphatidylserine serves as an electrostatic mediator as well as a buffer for bioactive fatty acids. Cardiolipin as a phosphatidylglycerol metabolite allows the maintenance of membrane potential of the inner mitochondrial membrane. These lipids are amphipathic to allow self-organization.
- The packing of membranes is facilitated by sphingolipids that control the permeability and fluidity. Sphingolipids are components of lipid rafts that have been discussed above.
- Sphingolipids and cholesterol are associated with lipid rafts.
- Lipids also function as bioactive molecules in signaling: the brain has high levels of PUFAs; arachidonic acid and docosahexaenoic acid.
- The release of neurotransmitter is suppressed by the derivatives of arachidonic acid: anandamide, and 2-arachidonoylglycerol.

- Inflammation in the brain is suppressed by omega-3 fatty acids while omega-6 fatty acids exert pro-inflammatory effects.
- The risk of neuroinflammatory diseases is lowered by docosahexaenoic acid.
- 12% of the total lipid content is composed of lipid content.
- The role of lipid rafts in the brain ranges from the remodeling of actin/tubulin, transport of neurotransmitters, redox signaling as well as binding and trafficking (Tracey et al, 2018).

4.12 ALTERED LIPIDS IN ALZHEIMER'S DISEASE

Research points out at the changes and involvement of lipids in diseases that is suggestive of the use of lipids to serve as biomarkers:

- Susceptibility biomarker: to look at risks
- Monitoring biomarker: to monitor the course of a disease
- Predictive biomarker: to assess which patient will benefit from a treatment

Table 4.2: A summary of few lipid biomarkers reported in studies in cerebrospinal fluid and blood samples

Lipids	**Sample**	**Findings**
Phospholipids	Blood	The levels of PC, phosphatidylcholine: PC 16:0/20:5, PC 16:0/22:6, and PC 18:0/22:6 decreased in Alzheimer's disease and mild cognitive impairment against controls.
	Cerebrospinal fluid	The ratio of lysoPC and lysoPC/PC (PC, phosphatidylcholine) was lower in Alzheimer's disease and mild cognitive impairment against controls.
	Cerebrospinal fluid	The levels of PPE, plasmalogen phosphoethanolamine: PPE 34:2 and PPE 36:4 were increased while that of PC 34:6, PC 36:5, and PC 40:6 were lowered in autosomal-dominant AD (ADAD) against controls

Sphingolipids	Cerebrospinal fluid	The levels of ceramides were increased in AD and ALS against controls.
	Cerebrospinal fluid	Sulfatide was decreased by 40% in dementia.
	Blood	Ceramide levels of C16:0 and C24:0 were linked to increased risks of AD in Women's Health Aging Study II.

(Wong et al, 2017)

4.13 LIPIDS AND DRUG DELIVERY

Lipid vesicles called as liposomes are small scale (nano or micro) structures composed of lipid bilayers with an aqueous core. These have been employed in delivery and trafficking of nucleic acids/ proteins and drugs: the year 2011 saw the generation of more than $750 million from this industry with 13 clinically approved liposome agents. The use of appropriate lipids (natural or designed to contain linkers or groups) can allow the delivery of cargo. The *in vivo* behavior of liposomes depends on the process of assembly and the choice of lipids or more specifically the physicochemical interactions of the different lipids.

Natural lipids used include phosphatidylcholine (PC), phosphatidylethanolamine (PE), sphingomyelin, cardiolipin and glycerides that have been used to make the liposomes Doxil, DepoCyt and AmBisome. The use of synthetic lipids allow for controlled release with extended time in the blood. The head groups of lipids can be designed such that the targeting of specific cell types is achieved.

The various aspects of liposomes are summarized in Figure 4.3:

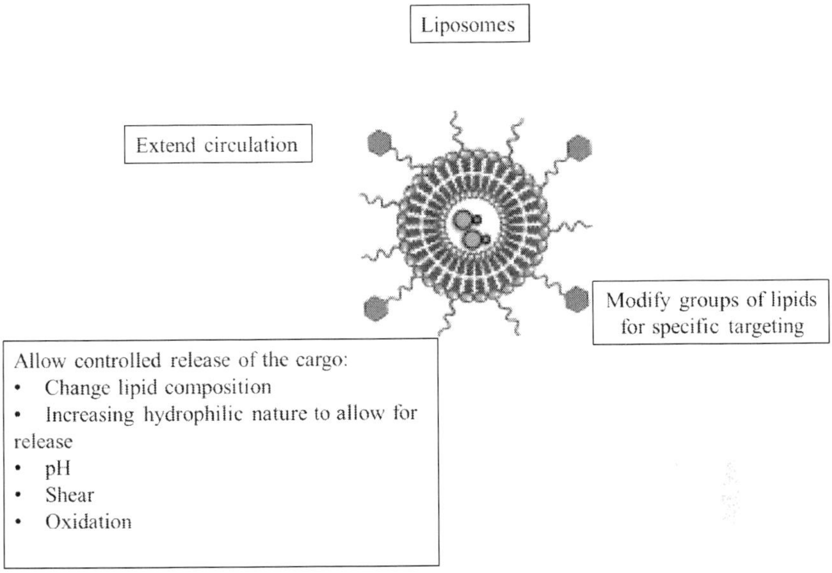

The distribution, safety, stability, and pharmacokinetics of small molecule drugs or prodrugs can be augmented by incorporating them into a liposome. Few examples include:

- An agent used in chemotherapy called MMC displays toxic effects; this has been addressed by the use of lipid prodrugs such as using a cleavable dithiobenzyl linker to conjugate MMC to 1, 2-distearoyl glycerol lipid that has shown promise in preclinical assays.
- The use of an inhibitor of topoisomerase II called Dox has received much attention in nanosystems. Conjugating albumin to doxorubicin yields AldoxorubicinR (CytRX) that is promising in early trials. The efficiency of Dox was increased by conjugating it with tocopherol succinate.

The formation of liposomes with multiple features has been facilitated by research to prolong the circulation and stability in the body as well as increased loading of drugs as well as targeted and controlled release. For example, the use of remote loading, increased hydrophilic nature of the cargo to allow its loading into the aqueous compartment as well as changing the formulation of the liposomes allow for controlled release. Another approach involves the use of shear or oxidation or pH changes to allow a burst of the liposome in the correct or desired environment. Increased research on the interactions among the lipid molecules in a liposome as well as increasing cargo stability can further augment this field of nanomedicine (Reviewed by Kohli et al, 2014).

4.14 FINAL NOTE

This chapter discussed the role or alterations seen in the levels or profiles of lipids in infectious diseases such as dengue and malaria as well as lysosomal diseases, and neurodegenerative diseases. Most of these studies have recorded significant changes in several crucial diseases such as kidney disease, cancer and neurodegeneration. Such changes can serve as indications of lipids to be used as biomarkers of disease.

The chapter also presented functional lipids or lipids that apart from nutrition also aid in preventing disease or promoting health as shown by research. The role of lipids in drug delivery agents or liposomes has also been covered in order to address their potential in the treatment of several diseases.

4.15 REFERENCES

1. Alabdulkarim B, Bakeet ZAN, Arzoo S. Role of some functional lipids in preventing diseases and promoting health. *Journal of King Saud University – Science.* 2012;24:319–329.
2. Bulbul MC, Dagel T, Afsar B. Disorders of Lipid Metabolism in Chronic Kidney Disease. *Blood Purif.* 2018;46:144–152.
3. Dias RM, Vieira JLF, da Conceição Cabral B. Lipid Profile of Children with Malaria by *Plasmodium vivax*. *J Trop Med.* 2016;2016:Article ID 9052612, 5 pages.
4. Dodge JC. Lipid Involvement in Neurodegenerative Diseases of the Motor System: Insights from Lysosomal Storage Diseases. *Front Mol Neurosci.* 2017;10:356. doi: 10.3389/fnmol.2017.00356
5. Fernandez C, Sandin M, Sampaio JL, Almgren P, Narkiewicz K, et al. Plasma Lipid Composition and Risk of Developing Cardiovascular Disease. *PLoS ONE.* 2013;8(8): e71846. doi:10.1371/journal.pone.0071846.
6. Finley JW, Klemann LP, Levelle GA, Otterburn MS, Walchak CG. Caloric Availability of Salatrium in Rats and Humans. *J Agric Food Chem.* 1994;42(2):495-499.
7. Gupta A, Paneri S, Lohokare R, Sarkar PD, Jain A. Clinico-biochemical correlation between psoriasis and lipid profile. *Int J Res Med Sci.* 2016 Jun;4(6):1966-1969.
8. Hamam F. Specialty Lipids in Health and Disease. *Food Nutr Sci.* 2013;4:63-70.
9. Kohli AG, Kierstead PH, Venditto VJ, et al. Designer lipids for drug delivery: from heads to tails. *J Control Release.* 2014 September 28;0:274–287.
10. Luo X, Cheng C, Tan Z, et al. Emerging roles of lipid metabolism in cancer metastasis. *Mol Cancer.* 2017; 16:76.
11. Melo CFOR, Delafiori J, Dabaja MZ, de Oliveira DN, et al. The role of lipids in the inception, maintenance and complications of dengue virus infection. *Scientific Reports.* 2018;8:11826.
12. Puri BK. Nutrients, Stress and Medical Disorder, Nutrition and Health: Lipids and Depression. Totowa, NJ: Human Press Inc.; 2005.
13. Tirinato L, Pagliari F, Limongi T. An Overview of Lipid Droplets in Cancer and Cancer Stem Cells. Hindawi. *Stem Cells Int.* 2017;2017:

Article ID 1656053, 17 pages.

14. Tracey TJ, Steyn FJ, Wolvetang EJ, Ngo ST. Neuronal Lipid Metabolism: Multiple Pathways Driving Functional Outcomes in Health and Disease. *Front Mol Neurosci*. 2018;11:10. Published 2018 Jan 23. doi:10.3389/fnmol.2018.00010

15. Wong MW, Braidy N, Poljak A, et al. Dysregulation of lipids in Alzheimer's disease and their role as potential biomarkers. *Alzheimer's & Dementia.* 2017;13:810-827.

CHAPTER 5

ALGAL AND VEGETABLE OILS AS SUSTAINABLE FISH OIL SUBSTITUTES IN RAINBOW TROUT DIETS: AN APPROACH TO REDUCE CONTAMINANT EXPOSURE

Amélie Bélanger-Lamonde[1] , Pallab K. Sarker[1] , Pierre Ayotte[2,3] , Janice L.Bailey[1] , Dominique P. Bureau[4] , P. Yvan Chouinard[1] , Éric Dewailly[3] , AlainLeblanc[2] , Jean-Philippe Weber[2] , and Grant W. Vandenberg[1]

[1]Département des Sciences Animales, Pavillon Paul-Comtois, Université Laval, Québec, QC, Canada G1K 7P4
[2]Centre de Toxicologie, Institut National de Santé Publique du Québec (INSPQ), Québec, QC, Canada G1V 5B3
[3]Axe Santé des Populations et Pratiques Optimales en Santé, Centre de Recherche du CHU de Québec, QC, Canada G1V 5B3
[4]Fish Nutrition Research Laboratory, Department of Animal and Poultry Science, University of Guelph, Guelph, ON, Canada N1G 2W1

Citation: Amélie Bélanger-Lamonde, Pallab K. Sarker, Pierre Ayotte, Janice L.Bailey, Dominique P. Bureau, P. Yvan Chouinard, Éric Dewailly, AlainLeblanc, Jean-Philippe Weber, and Grant W. Vandenberg. Algal and Vegetable Oils as Sustainable Fish Oil Substitutes in Rainbow Trout Diets: An Approach to Reduce Contaminant Exposure, Journal of Food Quality, Volume 2018, Article ID 7949782, 12 pages https://doi.org/10.1155/2018/7949782.

ABSTRACT

The aim of this study was to replace 75% of total fish oil (FO) and alter digestible protein/digestible energy (DP/DE) in rainbow trout feeds to minimize potentially hazardous environmental contaminant exposure. Two diets differing in DP/DE ratios (18 and 25) were combined with soybean (SO), canola (CO), and a blend of canola oil and Schizochytrium sp. (COS). Dietary lipids and DP/DE ratios did not affect apparent digestibility, growth, and somatic parameters. The n-3/n-6 levels decreased significantly in the growth trial, especially for the SO groups. A short washout trial restored n-3/n-6 levels for the CO and COS groups, irrespective of the DP/DE ratio, but not for the SO groups. At all sampling events, contaminant concentrations in fish flesh were lower than limits set by regulatory agencies in Canada, the US, and Europe. Contaminants were lower in the oil replacement diets compared to FO for toxaphenes, organochlorine pesticides, and PCBs but not flame retardants during the growth phase. At the end of the washout phase, no differences were detected. Thus, this study revealed that replacing 75% of total fish oil in rainbow trout feed by CO and COS, combined with a 25 DP/DE ratio, with a washout period seems to be the most efficient approach in terms of maximizing the total FO replacement and contribute to reducing POPs exposure.

INTRODUCTION

The contamination of aquaculture products by persistent organic pollutants (POPs) and heavy metals has become an issue of public concern, particularly after the publications reporting the contaminant levels of farmed salmonids [1, 2]. The main source of contaminants is from pelagic fish species used to produce fish meal and oil as ingredients in salmonid feeds [1–3]. To reduce the reliance on fish oils and address contaminant issues, efforts can be directed on reducing fish oil percentage in fish feed by partially replacing it with vegetable oils. Also, feeding diets having higher protein/lipid ratios at similar digestible energy levels can be a means to reduce the content of lipid-soluble contaminants. A number of studies have shown that a significant percentage of fish oil in salmonids feed can be replaced without affecting survival, growth, and feed efficiency [4–9]. However, vegetable oils have a different fatty acid profile than fish oils. When included at high dietary levels, they influence flesh fatty acid composition and reduce the concentration of omega-3 long-chain polyunsaturated fatty acids (n-3 LC-PUFA) such as docosahexaenoic acid (DHA) and eicosapentaenoic acid (EPA), well-known for their beneficial cardiovascular and cognitive properties [10]. Strategies to restore the levels n-3 LC-PUFA in farmed fish flesh include a washout period whereby fish oils are re-fed at higher levels following the growth period or the dietary inclusion of algal concentrates high in n-3 LC-PUFA. Indeed previous work has attempted to demonstrate the relevance of using a washout feed containing 100% fish oil to restore beneficial fatty acid levels with gilt-head seabream (Sparus aurata) [11, 12], Atlantic salmon (Salmo salar) [13, 14], and Nile tilapia (Oreochromis niloticus) [15]. Although promising, no study was able to sufficiently restore the n-3 LC-PUFA levels in the flesh because fish have a limited capacity of producing these fatty acids. However, by feeding vegetable oil-based feeds, combined with a high n-3 LC-PUFA source (Schizochytrium sp.), we hypothesize that initial n-3 LC-PUFA levels will be maintained so the washout period could be shorter than in previous studies. Bell et al. [16] concluded that replacing fish oil in the aquaculture feed by vegetable oils (ratio of 1 : 1 canola and linseed oils) plus a washout period is efficient in reducing persistent organic pollutants (POPs) accumulation in Atlantic salmon. Although vegetable oils are excellent candidates to replace marine oils because of their low contaminant concentrations, further investigation must be done to maintain beneficial n-3 LC-PUFA levels in the fillets.

The aim of this study was to determine the loading of several POPs and fatty acid profiles in flesh, when replacing 75% of total fish oil with a blend of vegetable oils and a high n-3 LC-PUFA algal source (Schizochytrium sp.), and altering the dietary digestible protein/energy (DP/DE) levels in rainbow trout (Oncorhynchus mykiss) feeds. A growth trial (9 months) was followed by a washout trial (3 months) to compare POPs and mercury concentrations, while evaluating the restoration of fatty acid profiles. During the growth period, we hypothesize that replacing fish oil with vegetable oils and modifying protein/lipid levels in fish feed will result in significantly lower contaminant levels in the flesh, while maximizing the utilization of lipids as a source of energy. For the washout period, we predict that feeding fish with a "finishing feed" containing high levels of n-3 LC-PUFA will permit restoration of these beneficial components in the flesh.

MATERIALS AND METHODS

Fish and Feeding

Rainbow trout (Oncorhynchus mykiss) weighing 43.6 ± 1.0 g at the beginning of growth trial were randomly allocated in 24 tanks of 340 L in a fresh water flow-through system at the Alma Aquaculture Research Station (AARS) in Elora (ON). Water temperature was a constant 8.5°C, incoming flow rate 12 l/min, and oxygen was maintained at >80% saturation. Photoperiod was set at a 12/12 light/dark cycle. Tanks were stocked with 45 fish per tank in order not to exceed a density of 60 kg/m^3 per tank at the end of the trial, based on known growth rates. Each diet was assigned to three replicate tanks (n = 3) in a completely randomized design to limit any environmental effects resulting from lighting intensity or floor traffic. Prior to the beginning of growth trial, fish were acclimated for two weeks. On a weekly basis, fish were hand-fed until apparent satiation two days and then at 75% satiation the next five days with belt feeders, twice a day. Pellet size varied according to fish body size, from 3 to 6 mm [17]. All aspects of the trial were approved by the Comité de Protection des Animaux de l'Université Laval.

Feeds

Growth Trial

Isoenergetic diets were formulated according to the NRC requirements for rainbow trout (1993). Two digestible protein/digestible energy ratios (18 and 25 DP/DE), based on those employed by Azevedo et al. [18], were formulated with four different sources of lipids. The lipid sources were herring oil (FO), soybean oil (SO), canola oil (CO), and a blend of canola oil and Schizochytrium sp. biomass (COS) (Table 1). Schizochytrium sp. biomass is a dried marine algae preparation rich in docosahexaenoic acid (DHA; 22:6w3). The COS blend was formulated to provide an equivalent level of DHA versus FO, while providing equivalent energy to other FO replacements via CO. The following diets were formulated: 18 DP/DE with 100% FO (18 FO); 18 DP/DE or replaced with 75% SO (18 SO); 18 DP/DE with 75% CO (18 CO); 18 DP/DE with 75% of a blend of CO and Schizochytrium sp. (18 COS); 25 DP/DE with 100% FO (25 FO); 25 DP/DE with 75% SO (25 SO); 25 DP/DE with 75% CO (25 CO); and 25 DP/DE with 75% of a blend of CO and Schizochytrium sp. (25 COS). Algae biomass was added to approximate n-3 LC-PUFA concentration in fish oil. Fatty acid profile of the experimental diets is shown in Table 2. Herring oil was purchased from Corey Aquafeeds (Fredericton, NB), soybean oil from Soya Excel (Beloeil, QC) and canola oil from Bunge (Montreal, QC). Schizochytrium sp. (S-Type Gold Fat) was supplied by Advanced BioNutrition (Columbia, MD). All dry ingredients were mixed, after which the mixture was blended with either fish oil or a blend of fish oil and the alternate source of lipid (10% of total oil was reserved for top-coating pellets). Pellets were steam pelleted with a California Pellet Mill (model CPM CL-5, California Laboratory Pellet Mill Co., Crawfordsville, IN), dried under forced air at room temperature for 24 h and then sieved. Diets were then coated with 10% of appropriate oil.

Table 1: Feed components and chemical composition of experimental diets (g/kg dry matter)

	18 FO	18 SO	18 CO	18 COS	25 FO	25 SO	25 CO	25 COS
Ingredient composition								
Fish oil, herring	21.3	2.7	2.7	2.8	13.5	0.0	0.0	0.0
Soybean oil	0.0	18.6	0.0	0.0	0.0	13.5	0.0	0.0
Canola oil	0.0	0.0	18.6	15.0	0.0	0.0	13.5	10.5
Schizochytrium sp.	0.0	0.0	0.0	9.0	0.0	0.0	0.0	6.5
Herring meal	22.0	22.0	22.0	22.0	32.5	32.5	32.5	32.5
Blood cell meal								
AP301	7.0	7.0	7.0	7.0	10.0	10.0	10.0	10.0
Corn gluten meal	21.0	21.0	21.0	20.0	34.0	34.0	34.0	33.0
Wheat middlings	13.0	13.0	13.0	10.5	4.0	4.0	4.0	2.3
Whey	11.4	11.4	11.4	9.5	2.8	2.8	2.8	2.0
Sipernat™	1.0	1.0	1.0	1.0	1.0	1.0	1.0	1.0
CaHPO4	0.5	0.5	0.5	0.5	0.5	0.5	0.5	0.5
Lysine	0.2	0.2	0.2	0.2	0.2	0.2	0.2	0.2
Vitamin/mineral premix[1]	0.6	0.6	0.6	0.6	0.6	0.6	0.6	0.6
Carophyll pink premix	0.3	0.3	0.3	0.3	0.3	0.3	0.3	0.3
Chemical composition								
Dry matter (g/kg)	92.0	92.0	91.2	92.8	92.5	93.1	92.6	92.0
Crude protein	45.2	44.6	45.4	46.4	61.6	61.7	62.1	61.7
Total lipid	21.6	21.4	21.2	22.5	15.3	15.4	15.9	16.2
Ash	7.5	7.5	7.7	8.1	8.5	8.4	8.4	8.9

Gross energy (MJ/kg DM)	24.4	24.3	24.4	24.5	24.1	23.9	24.1	24.0
Calculated P/E ratio (g DM/MJ)[2]	18.5	18.3	18.6	18.9	25.5	25.8	25.8	25.7

18 FO, 18 DP/DE diet with 100% fish oil; 18 SO, 18 DP/DE diet with 75% soybean oil; 18 CO, 18 DP/DE diet with 75% canola oil; 18 COS, 18 DP/DE diet with 75% blend of canola oil and Schizochytrium sp.; 25 FO, 25 DP/DE diet with 100% fish oil; 25 SO, 24 DP/DE diet with 75% soybean oil; 25 CO, 25 DP/DE diet with 75% canola oil; 25 COS, 25 DP/DE diet with 75% blend of canola oil and Schizochytrium sp.; DM, dry matter; DP/DE, digestible protein/digestible energy (lipids). Canola oil, soybean oil, and Schizochytrium sp algae were devoid of PCBs, pesticides, toxaphenes, and flame retardants. Herring meal was found to contain the following: PCBs: 6.5 ppb; pesticides: 2.45 ppb; toxaphenes: 0.41 ppb; flame retardants: 0 ppb. Herring oil was found to contain the following: PCBs: 386 ppb; pesticides: 147 ppb; toxaphenes: 46.8 ppb; flame retardants: 49.9 ppb. [1]Salmonid vitamin/trace mineral premix from Corey (Fredericton, NB). [2]P/E ratio = protein/energy ratio; calculation based on crude protein and gross energy values determined for the experimental diets.

Table 2: Fatty acid profile of experimental diets (% of total fatty acids)

	18 FO	18 SO	18 CO	18 COS	25 FO	25 SO	25 CO	25 COS
Fatty acid								
14:0	3.4	0.0	0.0	3.2	5.3	0.0	0.0	5.3
16:0	14.7	13.1	8.4	12.6	13.2	14.4	9.5	12.0
18:0	2.6	3.7	2.6	2.2	2.1	4.1	2.9	1.7
Total SFA[1]	20.7	16.8	11.0	17.9	20.6	18.5	12.4	20.6
16:1	0.0	0.0	0.0	1.6	6.2	0.0	1.8	6.3
18:1	36.3	18.3	51.4	37.8	12.0	19.0	49.7	11.2
20:1	0.0	2.7	3.3	2.8	12.9	0.0	1.8	15.3
22:1	0.0	4.4	4.3	3.3	25.0	0.0	0.0	29.3
24:1	0.0	0.0	0.0	0.0	1.9	0.0	0.0	1.9
Total MUFA	36.3	25.4	58.9	45.5	58.1	19.0	53.3	64.0
18:2n-6	16.7	47.0	20.0	14.8	8.9	47.5	21.0	7.4

20 : 2n-6	0.0	0.0	0.0	0.0	0.0	0.0	0.0	0.0
20 : 3n-6	0.0	0.0	0.0	0.0	0.0	0.0	0.0	0.0
20 : 4n-6	0.0	0.0	0.0	0.0	0.0	0.0	0.0	0.0
22 : 5n-6	4.2	0.0	0.0	4.0	0.0	0.0	0.0	0.0
Total n-6[2]	20.9	47.0	20.0	18.7	8.9	47.5	21.0	7.4
18 : 3n-3	4.2	7.2	6.1	4.5	0.0	7.0	5.8	0.0
20 : 5n-3	5.0	1.8	1.9	2.0	4.2	4.7	4.4	3.1
22 : 6n-3	13.0	1.8	2.1	11.3	6.2	3.3	3.1	5.0
Total n-3	22.1	10.9	10.1	17.8	10.4	15.1	13.3	8.1
Total PUFA	43.0	57.8	30.1	36.5	19.3	62.5	34.3	15.5
n-3/n-6	1.1	0.2	0.5	1.0	1.2	0.3	0.6	1.1

18 FO, 18 DP/DE diet with 100% fish oil; 18 SO, 18 DP/DE diet with 75% soybean oil; 18 CO, 18 DP/DE diet with 75% canola oil; 18 COS, 18 DP/DE diet with 75% blend of canola oil and Schizochytrium sp.; 25 FO, 25 DP/DE diet with 100% fish oil; 25 SO, 24 DP/DE diet with 75% soybean oil; 25 CO, 25 DP/DE diet with 75% canola oil; 25 COS, 25 DP/DE diet with 75% blend of canola oil and Schizochytrium sp.; DM, dry matter; DP/DE, digestible protein/digestible energy (lipids); TFA, total fatty acid; TL, total lipid; SFA, saturated fatty acid; MUFA, monounsaturated acid; PUFA, polyunsaturated fatty acid. [1]Includes 15 : 0, 17 : 0, 20 : 0, and 22 : 0. [2]Includes 18 : 3n-6.

Washout Period

To restore beneficial n-3 LC-PUFA levels in fillets, fish oil-based diets were fed for a three-month washout period. During this period, fish were fed diets containing 100% fish oil.

Sampling Procedures and Evaluation of Growth Parameters

Growth Trial

Fish were weighed at the beginning of the experiment and then every 28 days. At time 0 (beginning of the experiment), 15 fish were removed from the initial population. Thereafter, 5 fish per tank were sampled every 3 months at month 3 (84 days), month 6 (168 days), and month 9 (252 days). At each sampling event, fish were killed with an overdose of tricaine methanesulphonate (MS-222). Body, heart, liver, viscera, and gonads (if they were undergoing sexual maturation) were weighed. Fork length was measured, and color was evaluated with a SalmoFan™. Each fillet was taken from the carcass, and skin was removed and stored in a Ziploc bag at −20°C for further analysis.

The effects on growth were determined by evaluating growth performance indices, such as weight gain, thermal growth unit coefficient (TGC), feed conversion ratio (FCR), and survival percentage. Somatic parameters including condition factor (CF) and hepatosomatic index (HSI) were also calculated using the following formulae [5]:

$$\mathrm{CF} = 100 \times \frac{\text{body weight (g)}}{\text{length at fork (mm)}^3},$$

$$\mathrm{HSI} = 100 \times \frac{\text{liver weight (g)}}{\text{body weight (g)}},$$

$$\mathrm{FCR} = \frac{\text{feed intake (g)}}{\text{weight gain (g)}},$$

$$\mathrm{TGC} = 100 \times \frac{\left(\text{final body weight (g)}^{1/3} - \text{initial body weight (g)}^{1/3}\right)}{(\text{temperature} \times \text{days})}. \quad (1)$$

Washout Period

Five fish per tank were sampled every 28 days, at time 9 (252 days), 10 (280 days), 11 (308 days), and 12 (336 days). Measurements and sample procedure were same as for the growth period of the experiment.

Chemical Composition

Subsamples of 10 g wet weight/relative tank weight were weighed from each fillet of the same tank, and then pooled together. Samples were taken from the middle of the upper part of the fillet. A pool of different feed sizes was collected for the analyses of the feeds. Analyses for dry matter, ash, crude protein, total lipid, gross energy, and fatty acid profile were performed. Fillet dry matter was obtained by first drying the pools in a lyophilizator for seven days, then in a forced air oven at 105°C overnight. Sample weight was recorded before and after drying, followed by cooling in a desiccator [19]. Feed dry matter was obtained by drying only in a forced-air oven. Ash content was obtained by dry ashing in porcelain crucibles in a muffle furnace at 500°C overnight and expressed as dry weight [19]. Gross energy was performed by bomb calorimetry (Parr Instrument Company, Inc., Moline,

IL) and calculated as percentage of dry matter. Crude protein was evaluated using a LECO FP-2000 instrument (LECO Corporation, St. Joseph, MI), then a nitrogen (N) conversion factor of N × 6.25, expressed as dry weight.

Total lipid composition was performed with a Soxhlet HT-TECATOR® extractor (Soxtec System HT12, Foss Tecator AB; Hoganas, Sweden), the solvent being diethyl ether at 100°C. Total lipids were expressed as dry weight. To determine the fatty acid composition of the diets and fillets, the AOAC FAME procedure [20] was used. Samples were analysed at the Centre de Recherche en Biotechnologies Marines, Rimouski (QC), on a GC FID gas chromatograph from Hewlett-Packard model 5890 series II (Palo Alto, CA) equipped with a CP-Sil 88 capillary column (100 m × 0.25 mm). Oven temperature was held at 80°C for 1 min before being raised to 215°C (2°C min^{-1}) for 30 min and maintained for 98 min. Injector temperature was 220°C, and detector temperature was 230°C. Fatty acid peaks were identified, quantified, and the gas chromatograph calibrated using pure methyl ester standards (Nu Chek Prep; Elysian, MN). Individual components were identified by comparing retention times with those of the standards and quantified by calculating area under the curve with the ChemStation Rev: A 10.01 program (Agilent Technologies; Santa Clara, CA).

Digestibility Measurements

To determine the digestibility of the experimental diets, Sipernat 50™ (Jefo Nutrition, Inc., Saint-Hyacinthe, QC) was included as an inert marker of acid insoluble ash (AIA) to determine apparent nutrient digestibility coefficients (ADCs). The digestibility trial was performed with rainbow trout (222.6 ± 17.2 g) randomly allocated in 24 tanks (160 L) (n = 3) in a recirculating system at the Laboratoire de Recherche en Sciences Aquatiques of Université Laval (Quebec, QC). Tanks were filled with 20 fish, reaching approximately 28 kg/m^3biomass per tank. Water temperature was maintained at 12.6 ± 0.4°C, and other environmental parameters remained within limits recommended for salmonids [17].

Fish were acclimated three days to the feed and temperature before beginning of digestibility trial. Fish were hand-fed until apparent satiation twice daily (9 : 00 and 16 : 00). Feces were collected once a day for nine days via a modified Guelph system based on Cho et al. [21], before the morning meal. They were stored at −20°C until the end of the digestibility trial, and then thawed in a refrigerator at 4°C. Feces were centrifuged to remove excess water and freeze dried for seven days prior to analysis to determine ADC for the nutrients and energy of test and reference diets [21, 22]:

$$ADC = 1 - \left(\frac{F}{D} \times \frac{D_i}{F_i}\right), \quad (2)$$

where D = % nutrient (or kJ/g gross energy) of diet; F = % nutrient (or kJ/g gross energy) of feces; D_i = % indigestible indicator (AIA) of diet; and F_i = % indigestible indicator (AIA) of feces.

Contaminant Analyses

Levels of POPs and mercury were evaluated at times 0, 9 (252 days), and 12 (336 days). Each fillet from the same tank was pooled by taking a subsample of 2 g wet weight/relative tank weight from the median dorsal region of the fillet; tissue pools were then homogenized. Experimental diets were also analysed for contaminant content. PBCs, organochlorine pesticides, toxaphenes, flame retardants, and mercury concentrations were evaluated by the toxicology laboratory of the Institut National de Santé Publique du Québec (INSPQ) in Quebec City (QC). Tissue samples were enriched with internal standards, mixed with dichloromethane and chemically dried with sodium sulfate. Organohalogenated compounds were then extracted from the matrix by ultrasound sonification. A portion of the organic solvent was used to determine the lipid weight in the sample. The residual fraction was concentrated and purified by gel permeation chromatography and florisil treatment. Extracts were analysed by gas chromatography mass spectrometry (gas chromatograph: Agilent, #6890; mass detector: Agilent, #5973; network and automatic sampler with automatic injector: Agilent, #7683; Agilent Technologies, Mississauga, ON). Ions generated were measured after negative chemical ionization. Analyte concentrations were evaluated by considering the % recovery of labeled internal standards (Agilent MSD CHEM #G1701CA; Agilent Technologies, Missisauga, ON). Mercury was analysed by cold vapour atomic absorption spectrometry (Mercury Monitor Model 100, Pharmacia; Piscataway, NJ) with an application range from 0.05 µg/g (detection limit to 10 µg/g for a sample weight of 0.5 g of wet tissue). Tissues were digested using concentrated nitric acid in pressurized Teflon vessels. An aliquot of the digest was then introduced into the system's reaction chamber (containing a reducing solution of cadmium chloride and stannous chloride). The mercury vapour was generated and detected followed by, aqueous calibration.

For PCBs, 14 congeners were analysed: 28, 52, 99, 101, 105, 118, 128, 138, 153, 156, 170, 180, 183, and 187. The following organochlorine pesticides were analysed: β-HCH, α-chlordane, γ-chlordane, cis-Nonachlor, p,p'-DDE, p,p'-DDT, hexachlorobenzene, mirex, oxychlordane, and trans-Nonachlor. Five toxaphene congeners were analysed: parlar no. 26 (T2), parlar no. 32, parlar no. 50 (T12), parlar no. 62 (T20), and parlar no. 69. PBDE 47, PBDE 99, PBDE 100, PBDE 153, and PBB 153 were the flame retardants analysed.

Statistical Analyses

Experimental values for growth performance, chemical composition, and fatty acid profile for the growth and washout periods were compared, as well as contaminant exposure and digestibility, via analysis of variance (ANOVA) in a complete randomized design, with each tank being the experimental unit (3 replicates per treatment combinations of DP/DE ratio and source of lipid). Values expressed in percentage were transformed with arcsin prior to the analysis. Normality was evaluated with Kolmogorov–Smirnov test. Differences between means were evaluated for significant differences with Tukey's multiple range test. The significance level was set at $p<0.05$. Statistical analyses were performed using SAS 8.0 (SAS Institute, Inc., Cary, NC).

RESULTS AND DISCUSSION

Growth Period

The 18 DP/DE diet had a greater percentage of oil and contained less fish meal than the 25 DP/DE diets. While the level of fish meal was not altered in vegetable oil-based diets, total fish oil was reduced by 75%. However, these diets must contribute in maintaining growth and somatic parameters, as well as the high quality of the flesh in terms of fatty acid content. The average fish mass increased continuously through the experiment without mortality, and no effects of dietary treatments on weight gain, TGC, condition factor, and hepatosomatic index were observed in this study (Table 3). Diet 18 FO had a significantly higher FCR (1.19), while 25 COS was lower than other experimental diets (0.98). Growth performance and somatic parameter data in this study indicate that soybean and canola oils (with or without the algal biomass) can replace 75% of total fish oil in rainbow trout feed, without affecting growth performance and survival.

Table 3: Effect of fish oil replacement with soybean oil, canola oil, and a blend of canola oil and Schizochytrium sp. biomass on rainbow trout growth, nutritive utilization, and somatic parameters after growth trial

	18 FO	18 SO	18 CO	18 COS	25 FO	25 SO	25 CO	25 COS	Pooled SEM
Initial weight (g)	43.6	44.1	44.0	43.2	44.0	43.3	43.2	43.7	0.2
Final weight (g)	611.7	591.7	570.2	583.0	591.0	572.3	535.7	574.1	6.5
Weight gain (g)	568.1	547.6	526.2	539.7	547.0	529.1	492.5	530.4	6.4
TGC	0.23	0.23	0.22	0.23	0.23	0.22	0.21	0.22	0.00
FCR	1.19^{a}	$1.06^{b,c,d}$	$1.14^{a,b}$	$1.09^{b,c}$	$1.07^{b,c,d}$	$1.02^{c,d}$	$1.05^{b,c,d}$	0.98^{d}	0.01
CF	1.61	1.61	1.59	1.62	1.49	1.47	1.58	1.51	0.02
HSI	1.28	1.12	1.15	1.12	1.18	1.24	1.28	1.21	0.02
Survival (%)	97.8	100.0	100.0	99.3	100.0	98.5	99.3	98.5	0.1

18 FO, 18 DP/DE diet with 100% fish oil; 18 SO, 18 DP/DE diet with 75% soybean oil; 18 CO, 18 DP/DE diet with 75% canola oil; 18 COS, 18 DP/DE diet with 75% blend of canola oil and Schizochytrium sp.; 25 FO, 25 DP/DE diet with 100% fish oil; 25 SO, 25 DP/DE diet with 75% soybean oil; 25 CO, 25 DP/DE diet with 75% canola oil; 25 COS, 25 DP/DE diet with 75% blend of canola oil and Schizochytrium sp.; DM, dry matter; DP/DE, digestible protein/digestible energy (lipids); TGC, thermal growth coefficient; FCR, feed conversion ratio; CF, condition factor; HSI, hepatosomatic index. Data represent the mean ± standard error means of three replicates, except for survival data presented as mean ± standard deviation of three replicates. Within the same line, different superscripts indicate significant differences due to the diet ($p<0.05$).

With regards to fillet composition, no significant differences were observed for dry matter and ash content among the dietary groups tested. However, crude protein, total lipid, gross energy, and fatty acid composition were significantly different between DP/DE ratios and dietary lipid sources (Table 4). As the composition of dietary fatty acids has an effect on the muscle fatty acids, the high accumulation of 18 : 2n-6 in the flesh, especially the SO groups, could be due to the direct absorption of dietary fatty acids in the muscle [23, 24]. Moreover, these percentages could be explained by the great affinity of 18 : 2n-6 and the acyltransferases that synthesize phospholipids containing this fatty acid [5]. Also, the accumulation 20 : 3n-6, the product of the Δ6 desaturation and elongation of 18 : 2n-6, reveals a high activity of these enzymes when fish are fed high percentages of soybean oil [25]. With Atlantic salmon fed high levels of sunflower oil, Bell et al. [26] observed similar comparisons in macrophage phospholipids. When comparing the 18:2n-6 availability in the diets with the percentage in the fillets, an important decrease was observed. With the exception of the SO groups, no metabolic intermediates were found for the desaturation and elongation of 18:2n-6 to 22:5n-6 suggesting that 18:2n-6 was necessary for other physiologic requirements, for example, catabolizing energy [27, 28]. The 22:5n-6 in the COS groups appears to come from the direct absorption of DPA from the algal biomass.

Table 4: Effects of fish oil replacement with soybean oil, canola oil, and a blend of canola and Schizochytrium sp. biomass on fillet composition (g/kg DM) and fatty acid profile (% of total fatty acids) after growth period

	18 FO	18 SO	18 CO	18 COS	25 FO	25 SO	25 CO	25 COS	Pooled SEM
Fatty acid									
14:0	4.2^{a}	0.0^{c}	0.0^{c}	2.2^{a,b}	3.7^{a}	0.5^{b,c}	0.5^{b,c}	2.2^{a,b}	0.3
16:0	16.5^{b,c,d}	16.3^{c,d}	13.0^{e}	14.9^{d}	18.3^{a,b}	19.0^{a}	16.6^{b,c}	17.2^{a,b,c}	0.4
18:0	3.0^{f}	4.7^{b}	3.9^{c,d}	3.4^{e,f}	3.7^{d,e}	5.5^{a}	4.6^{b}	4.3^{b,c}	0.2
Total SFA[1]	23.8^{a}	21.0^{b}	16.8^{c}	20.5^{b}	25.7^{a}	25.0^{a}	21.8^{b}	23.8^{a}	0.6
16:1	6.7^{a}	1.2^{c,d}	1.7^{b,c}	0.6^{d}	6.6^{a}	2.6^{b}	3.0^{b}	2.6^{b}	0.5
18:1	17.4^{c}	18.2^{c}	43.5^{a}	34.8^{b}	20.8^{c}	20.7^{c}	40.5^{a}	34.6^{b}	2.1
20:1	12.2^{a}	2.4^{b}	3.8^{b}	3.2^{b}	9.3^{a}	0.4^{c}	2.5^{b}	2.3^{b}	0.8
22:1	15.9^{a}	1.7^{c}	0.0^{d}	2.6^{c}	11.2^{b}	0.0^{d}	0.0^{d}	0.0^{d}	1.1
24:1	2.0^{a}	0.0^{c}	0.0^{c}	0.0^{c}	1.8^{b}	0.0^{c}	0.0^{c}	0.0^{c}	0.2
Total MUFA	54.1^{a}	23.5^{e}	49.0^{a,b}	41.1^{c,d}	49.6^{a,b}	23.7^{e}	46.0^{b,c}	39.5^{d}	2.3
18:2n-6	5.5^{e}	32.0^{a}	13.6^{c}	11.3^{d}	6.0^{e}	26.2^{b}	12.3^{c,d}	11.0^{d}	1.9
20:2n-6	0.0^{b}	2.0^{a}	0.0^{b}	0.0^{b}	0.0^{b}	2.0^{a}	0.0^{b}	0.0^{b}	0.2
20:3n-6	0.0^{b}	2.3^{a}	0.0^{b}	0.0^{b}	0.0^{b}	2.3^{a}	0.5^{b}	0.0^{b}	0.2
20:4n-6	0.0^{b}	1.3^{a}	0.0^{b}	0.0^{b}	0.0^{b}	2.0^{a}	0.0^{b}	0.0^{b}	0.2
22:5n-6	0.0^{c}	0.0^{c}	0.0^{c}	3.7^{a}	0.0^{c}	0.0^{c}	0.0^{c}	3.1^{b}	0.3
Total n-6[2]	5.5^{e}	37.5^{a}	13.6^{c,d}	15.0^{c}	6.0^{e}	32.4^{b}	12.9^{d}	14.1^{c,d}	2.3
18:3n-3	0.0^{d}	3.8^{a}	3.5^{a}	3.1^{b}	0.0^{d}	2.9^{b}	2.6^{c}	2.4^{c}	0.3
20:5n-3	2.5^{a,b,c}	2.1^{b,c}	4.2^{a}	1.2^{c}	2.7^{a,b}	2.5^{a,b}	2.7^{a,b}	2.7^{a,b}	0.2
22:6n-3	14.2^{b,c}	12.1^{c}	12.8^{c}	19.1^{a}	16.0^{a,b,c}	13.4^{b,c}	14.1^{b,c}	17.6^{a,b}	0.6
Total n-3	16.7^{c}	17.9^{b,c}	20.5^{a,b,c}	23.3^{a}	18.7^{a,b,c}	18.8^{a,b,c}	19.4^{a,b,c}	22.7^{a,b}	0.5
Total PUFA	22.2^{d}	55.5^{a}	34.2^{b,c}	38.4^{b}	24.7^{d}	51.3^{a}	32.2^{c}	36.8^{b,c}	2.3

n-3/n-6	3.1[a]	0.5[c]	1.5[b]	1.6[b]	3.1[a]	0.6[c]	1.5[b]	1.6[b]	0.2
Chemical composition									
Dry matter (g/kg)	25.5	26.6	25.1	26.0	25.0	24.4	26.1	26.3	0.4
Crude protein	86.2[b]	88.6[a,b]	89.0[a,b]	88.7[a,b]	89.0[a,b]	91.1[a]	90.2[a]	89.0[a,b]	0.4
Total lipid	10.6[a]	8.1[a,b]	7.8[a,b]	8.6[a,b]	7.7[a,b]	5.9[b]	6.9[a,b]	8.0[a,b]	0.4
Ash	5.7	5.8	5.8	6.0	5.7	6.0	5.8	5.8	0.0
Gross energy (MJ/kg DM)	23.8[a]	23.3[a,b]	23.3[a,b]	23.3[a,b]	23.2[a,b]	22.8[b]	23.0[a,b]	23.3[a,b]	0.1

Data represent the mean ± standard error means of three replicates. Within the same line, different superscripts indicate significant differences ($p<0.05$). [1]Includes 15:0, 17:0, 20:0, and 22:0. [2]Includes 18:3n-6.

It seems no competition was present between 18:3n-3 and 18:2n-6 for the desaturation and elongation by the Δ6 desaturases, even when dietary 18:2n-6 levels were high (32.0% for 18 SO and 26.2% for 25 SO, Table 5). Similar results have been recorded for Atlantic salmon fed sunflower oil [26] and rainbow trout fed different vegetable oils [5]. Bell et al. [26] noted, however, a competition between these fatty acids when the diet was rich in 18:3n-3 (31.4%) from linseed oil, resulting in an inhibition of the bioconversion of 18:2n-6 to 22:5n-6.

Table 5: Effects of feeding with a washout diet of herring oil on fillet composition (g/kg DM) and fatty acid profile (% of total fatty acids)

	18 FO	18 SO	18 CO	18 COS	25 FO	25 SO	25 CO	25 COS	Pooled SEM
Fatty acid									
14:0	4.4[a]	3.1[b]	3.3[b]	3.3[b]	3.4[a,b]	2.8[b]	2.8[b]	3.0[b]	0.1
16:0	17.1	17.5	16.6	15.8	18.0	19.7	18.5	19.3	0.4
18:0	3.1[c]	4.1[a,b]	3.5[b,c]	3.2[c]	3.7[b,c]	4.8[a]	4.3[a,b]	4.3[a,b]	0.1
Total SFA[1]	24.6	24.7	23.4	22.3	25.2	27.3	25.6	26.6	0.5
16:1	6.1[a]	3.7[b,c]	4.3[a,b,c]	3.0[c]	5.0[a,b]	4.1[b,c]	4.2[b,c]	3.9[b,c]	0.3
18:1	18.4[c]	20.0[c]	30.6[a,b]	27.8[a,b,c]	26.8[a,b,c]	21.8[b,c]	31.8[a]	26.1[a,b,c]	1.3
20:1	11.5[a]	6.7[b,c]	7.7[b]	6.4[b,c]	7.5[b]	5.2[c]	5.4[c]	5.2[c]	0.5
22:1	15.1[a]	8.5[b]	9.2[b]	7.8[b,c]	8.7[b]	5.7[c,d]	5.3[c,d]	5.2[d]	0.7
24:1	1.9	0.8	1.1	0.0	0.8	0.0	0.0	0.0	0.2

Total MUFA	53.0^{a}	$36.9^{c,d}$	53.0^{a}	$44.9^{a,b,c,d}$	$48.8^{a,b}$	36.8^{d}		$40.5^{b,c,d}$	1.3
18 : 2n-6	5.5^{b}	18.2^{a}	8.0^{b}	8.9^{b}	7.6^{b}	17.2^{a}	9.7^{b}	8.1^{b}	1.0
20 : 2n-6	0.0	0.0	0.0	0.0	0.0	0.0	0.0	0.0	0.0
20 : 3n-6	0.0	0.0	0.0	0.0	0.0	0.0	0.0	0.0	0.0
20 : 4n-6	0.0	0.0	0.0	0.0	0.0	0.0	0.0	0.0	0.0
22 : 5n-6	0.0^{b}	0.0^{b}	0.0^{b}	2.2^{a}	0.0^{b}	0.0^{b}	0.0^{b}	2.1^{a}	0.2
Total n-6[2]	5.5^{d}	18.2^{a}	$8.0^{c,d}$	$11.1^{b,c}$	$7.6^{c,d}$	$17.2^{a,b}$	$9.7^{c,d}$	$10.2^{c,d}$	1.0
18 : 3n-3	0.0	2.2	1.3	2.0	1.0	1.9	1.2	0.0	0.2
20 : 5n-3	2.3	2.1	2.4	1.7	2.5	2.2	2.1	2.5	0.1
22 : 6n-3	14.6	13.1	15.4	18.0	14.9	14.6	14.7	20.2	0.6
Total n-3	16.8	17.3	19.1	21.7	18.4	18.7	18.0	22.7	0.6
Total PUFA	22.4^{c}	35.6^{a}	$27.2^{b,c}$	$32.8^{a,b}$	$26.0^{b,c}$	35.9^{a}	$27.7^{b,c}$	$32.9^{a,b}$	1.1
n-3/n-6	3.0^{a}	1.0^{b}	$2.4^{a,b}$	$2.0^{a,b}$	$2.4^{a,b}$	1.1^{b}	$1.8^{a,b}$	$2.2^{a,b}$	0.2
Chemical composition									
Dry matter (g/kg)	26.6	26.1	25.6	26.1	25.2	25.5	25.5	25.0	0.1
Crude protein	88.0	87.7	90.2	92.8	89.3	90.4	90.1	93.0	0.6
Total lipid	12.0	10.5	9.3	10.1	8.8	8.0	8.8	6.2	0.5
Ash	5.8	5.5	5.3	5.6	5.3	5.7	5.2	5.6	0.1
Gross energy (MJ/kg DM)	24.4	23.7	23.9	23.9	23.6	23.6	23.6	23.3	0.1

Data represent the mean ± standard error means of duplicates replicates. Within the same line, different superscripts indicate significant differences ($p<0.05$). [1]Includes 15 : 0, 17 : 0, 20 : 0, and 22 : 0. [2]Includes 18 : 3n-6.

Results of this study reveal that the composition of the flesh reflects a selective incorporation of essential fatty acids (EFAs), especially for EPA and DHA. These EFAs are probably adjusted to a narrowly defined physiological level into triacylglycerols and phospholipids [28–30]. In this regard, the preferential retention of EFA seems to influence the fatty acid content of the muscle [5]. Therefore DHA, the major PUFA in phospholipids, is affected to a lesser extent by the DHA content in the diet. Similar observations have been reported for brown trout (Salmo trutta L.) fed canola oil, poultry fat, pork lard, and oleine oil (Turchini et al., 2003). Elongation and desaturation of EFA was observed for all experimental groups, mainly for fish fed lower n-3 diets, and levels of n-3/n-6 were higher in the muscle than in the diet. In comparison with results from Greene and Selivonchick [28] where conversion of 18:3n-3 in EPA and DHA reached its limit at 14% of total muscle lipids, it seems that our experimental groups reached their limit at 20.3% (18 COS and 25 COS), irrespective of the total dietary n-3 available. Regardless of the initial content of EPA in the diets, experimental groups were able to accumulate and/or produce equivalent percentages of EPA versus the FO diets. The 18 CO groups even had a higher content of EPA than its control FO group. DHA accumulation in the muscle also followed this trend, and the overall muscle composition in DHA tended to be similar than for fish oil-fed trout. With this capacity of synthesizing DHA to maintain its concentration in the flesh, fish demonstrate the importance of this structural lipid for its own physiological requirements [23, 30]. In addition, fish fed with a blend of canola and algae had the greatest accumulation of DHA. When comparing CO and COS groups, it is clear that Schizochytrium sp. contributes to the overall DHA content in the muscles. As a result, total n-3 fatty acids were maintained or higher than for the control FO groups.

It is now well documented that, in terms of human health, it is essential to increase the daily consumption of n-3, mainly EPA and DHA [31] and to reduce n-6 daily intake, particularly 18:2n-6 [32]. For the growth phase alone, however, n-3/n-6 ratios were too low compared with the FO groups because of the high content of n-6 in the flesh. Thus, canola and the blend of canola and algae were good-quality candidates for 75% fish oil replacement in rainbow trout feeds, irrespective of the DP/DE ratio used. However, a washout phase is required to balance beneficial n-3 levels in the final product.

Washout Period

With a switch to a diet comprised of 100% fish oil, TGC values (mean 0.16) dropped about 27% between growth and washout trials (Table 6). As fish grow, their potential for rapid growth diminishes, because nutrient utilization is aimed for physiological changes that come with maturity, for example reproduction and maintenance [33–35]. This shift in nutrient partitioning could be responsible for the decrease of TGC values in the washout phase because individuals were larger than in the growth phase, and some were already mature. FCR values (mean 1.23) were similar to 50% (1.25) and 100% (1.18) rapeseed oil feeds observed by Bell et al. [16].

Table 6: Effect of feeding with a washout diet of herring oil on rainbow trout growth, nutritive utilization, and somatic parameters

	18 FO	18 SO	18 CO	18 COS	25 FO	25 SO	25 CO	25 COS	Pooled SEM
Initial weight (g)	617.1	596.9	570.2	577.8	580.1	561.3	535.7	549.7	7.7
Final weight (g)	916.6	949.8	906.4	938.7	815.8	776.2	797.5	782.4	19.0
Weight gain (g)	299.5	352.9	336.2	220.7	235.7	214.9	261.7	232.8	14.9
TGC	0.17	0.20	0.19	0.13	0.14	0.13	0.16	0.14	0.01
FCR	1.33	1.16	1.23	1.29	1.16	1.17	1.29	1.19	0.03
CF	1.58[a,b]	1.64[a,b]	1.65[a]	1.55[a,b]	1.58[a,b]	1.47[b]	1.64[a,b]	1.58[a,b]	0.02
HSI	1.06	1.22	1.16	1.17	1.22	1.11	1.21	1.26	0.02
Survival (%)	100.0	100.0	100.0	100.0	100.0	100.0	100.0	100.0	0.0

18 FO, 18 DP/DE diet with 100% fish oil; 18 SO, 18 DP/DE diet with 75% soybean oil; 18 CO, 18 DP/DE diet with 75% canola oil; 18 COS, 18 DP/DE diet with 75% blend of canola oil and Schizochytrium sp.; 24 FO, 24 DP/DE diet with 100% fish oil; 25 SO, 25 DP/DE diet with 75% soybean oil; 25 CO, 25 DP/DE diet with 75% canola oil; 25 COS, 25 DP/DE diet with 75% blend of canola oil and Schizochytrium sp.; DM, dry matter; DP/DE, digestible protein/digestible energy (lipids); TGC, thermal growth coefficient; FCR, feed conversion ratio; CF, condition factor; HSI, hepatosomatic index. Data represent the mean ± standard error means of duplicates, except for survival data presented as mean ± standard deviation of duplicates. Within the same line, different superscripts indicate significant differences due to the diet ($p<0.05$).

In this study, the washout period represented 33% of the growth period, which is important when determining the time required to restore beneficial fatty acid levels in rainbow trout. The aim is to shorten the washout period to maintain low POP levels and minimize costly fish by-product use in the whole production cycle, while allowing EFA restoration. The DP/DE ratios did not significantly influence the fatty acid accumulation in the flesh for the 25 DP/DE groups (Table 5). In the 18 DP/DE groups, significant differences were observed between the FO and the SO groups. Whereas the flesh n-3 profile was equivalent across all diet treatments after the washout period, and the n-3/n-6 ratios differed. Even after a washout period, fish previously fed with soybean oil demonstrated elevated n-6 fatty acid levels; therefore, the inclusion of this oil in rainbow trout feed should be questioned, in terms of desirable n-3/n-6 ratios profiles. In the growth phase, we observed that trout fed soybean oil accumulated high levels of n-6 in the flesh, reducing the quality for human consumption. During the washout phase, fish previously fed with vegetable oils, particularly soybean oil, significantly reduced their 18:2n-6 content, consequently, total n-6 levels were lowered. For the SO groups, this reduction of n-6 concentration was around 50%. For the other groups, the switch to lower n-6 fatty acids content was less apparent because of reduced n-6 accumulation during the growth phase than the SO groups. Irrespective of the DP/DE ratio, fish from the CO and COS groups restored n-6 levels, but the SO groups still had higher levels in the flesh. A longer washout period could have contributed in the total restoration of beneficial fatty acids for the SO groups, but it also could have raised the POP levels.

While n-6 levels were decreased in the washout period, n-3 levels remained as they were in the growth period. 18:3n-3 levels were lowered for all vegetable oil-fed groups, but DHA levels were increased, balancing the total n-3 content. EPA percentages remained the same for overall diets. For the 18 CO group, however, EPA levels significantly fell to reach approximate final 18 FO levels. In summary, n-3 levels were restored in all experimental groups, although a tendency of higher accumulation of DHA in the COS groups can be distinguished. The washout trial significantly reduced PUFA levels by lowering total n-6 and had a direct effect to increase n-3/n-6 ratios. Trout from the CO and COS groups restored their fatty acid levels, but the SO groups accumulated a higher n-6 level. Further analysis would be needed to understand the restoration profile of fatty acids to determine the shortest washout period required to restore beneficial fatty acids levels. In addition, after a short washout period that contributes in restoring beneficial EFA levels, canola oil and the blend of canola oil plus Schizochytrium sp. are high-quality candidates for replacing 75% of fish oil in rainbow trout diet, irrespective of the DP/DE ratio in the diets.

Contaminant Loading

Levels of POPs in flesh were well under recommendations for mercury (1 ppm), PCBs (2.0 ppm), toxaphenes (1.0 ppm), and organochlorine pesticides (5.0 ppm) (European Committee, 2007) [36, 37]. The majority of contamination exposure comes from two ingredients in the feed: fish meal and to a larger extent, fish oil (Table 1). None of the vegetable oils contained POPs or mercury, nor did the algal biomass. Of all feeds, the 18 FO diet had the highest concentration of fish oil and displayed the highest levels of PCBs, organochlorine pesticides, and toxaphenes (Figure 1). Bell et al.

[16] replaced 17% and 35% fish oil by a 1 : 1 combination of linseed and canola oil fed to Atlantic salmon. The growth period was of 115 weeks; afterwards a washout feed of 35% fish oil was given for 24 weeks. The dioxin and dioxin-like PCB concentrations in the flesh followed the pattern of the present study, with bioaccumulation being greater in fish feed high percentages of fish oil. More recent research demonstrated that Atlantic salmon produced using diets with 75% of the supplemental anchovy oil (61% of total dietary lipid) replaced by cold-pressed flaxseed oil had 61% lower levels of PCBs compared to fish grown on diets with 100% fish oil without compromising growth performance [38]. In this respect, the replacement of fish oil by vegetable oils is an excellent alternative for reducing contaminant exposure for rainbow trout. Since these oils alter the quality of the flesh by increasing n-6 and n-9 levels while diminishing n-3 levels, a washout feed should be considered to restore beneficial n-3 LC-PUFA levels in the flesh.

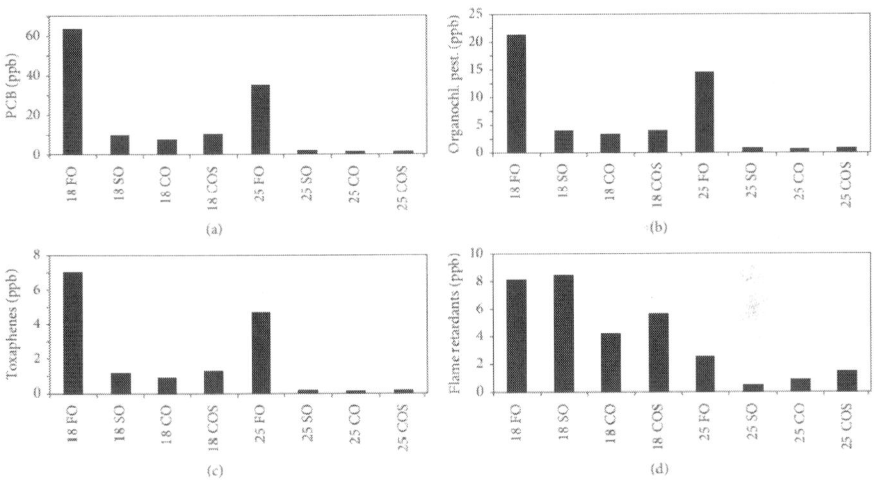

Figure 1: Concentrations of (a) PCBs (ppb), (b) organochlorine pesticides (ppb), (c) toxaphenes (ppb), and (d) flame retardants (ppb) in experimental diets. 18 FO, 18 DP/DE diet with 100% fish oil; 18 SO, 18 DP/DE diet with 75% soybean oil; 18 CO, 18 DP/DE diet with 75% canola oil; 18 COS, 18 DP/DE diet with 75% blend of canola oil and Schizochytrium sp.; 25 FO, 25 DP/DE diet with 100% fish oil; 25 SO, 25 DP/DE diet with 75% soybean oil; 25 CO, 25 DP/DE diet with 75% canola oil; 25 COS, 25 DP/DE diet with 75% blend of canola oil and Schizochytrium sp.; DM, dry matter; DP/DE, digestible protein/ digestible energy (lipids).

POP levels for initial fillets, growth, and washout phase fillets are shown in Figure 2. In growth trial fillets, PCBs were significantly higher for 18

FO than other groups and were the lowest for the 25 SO, CO, COS groups (Figure 2(a)). Organochlorine pesticides were highest in the 18 FO group and the lowest in the 25 SO and CO groups (Figure 2(b)). For the toxaphenes levels, the highest levels were found in the 18 FO and 25 FO groups and were lower in all other experimental diets (Figure 2(c)). Flame retardants were not significantly different for the growth trial (Figure 2(d)). Results from the growth trial in Figure 2 outline the greater accumulation of POPs in the FO groups, particularly for the 18 FO-fed fish. Vegetable oil sources contributed in reducing the POP levels, irrespective of the DP/DE ratios.

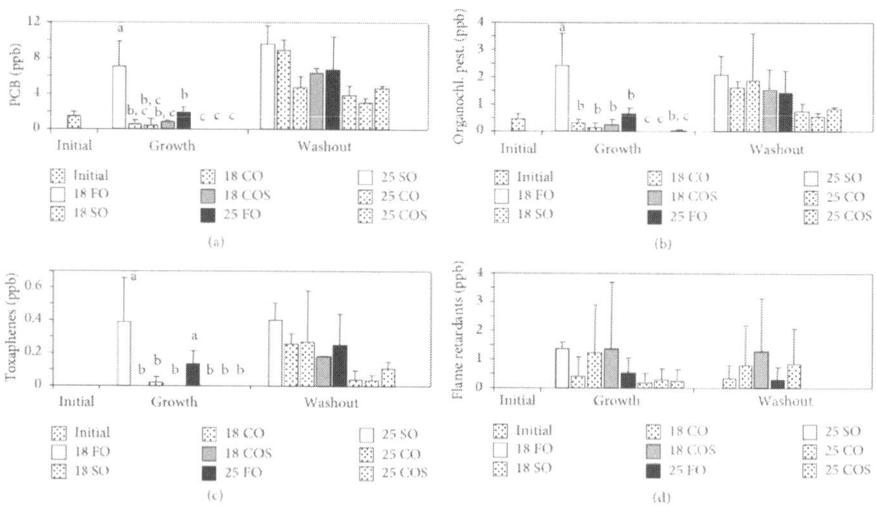

Figure 2: Concentrations of (a) PCBs (ppb), (b) organochlorine pesticides (ppb), (c) toxaphenes (ppb), and (d) flame retardants (ppb) in initial fish, at the end of growth period, and at the end of washout period. 18 FO, 18 DP/DE diet with 100% fish oil; 18 SO, 18 DP/DE diet with 75% soybean oil; 18 CO, 18 DP/DE diet with 75% canola oil; 18 COS, 18 DP/DE diet with 75% blend of canola oil and Schizochytrium sp.; 25 FO, 25 DP/DE diet with 100% fish oil; 25 SO, 25 DP/DE diet with 75% soybean oil; 25 CO, 25 DP/DE diet with 75% canola oil; 25 COS, 25 DP/DE diet with 75% blend of canola oil and Schizochytrium sp.; DM, dry matter; DP/DE, digestible protein/digestible energy (lipids). Data represent the mean of three replicates for initial data and growth trial; duplicates for washout trial. The error bars represent standard errors. Within the growth period, different superscripts indicate significant differences ($p<0.05$).

When comparing experimental results for the washout phase, a number of tendencies were noted; a switch to a 100% fish oil feed increased the accumulation of PCBs, organochlorine pesticides, and toxaphenes in the flesh. This statement is more obvious with the vegetable oil-feed groups, where POPs were accumulated to a larger extent, but still lower than in the fish oil-feed groups. Once again, a tendency of lower POP concentration is noticeable for the 25 DP/DE group, but is not significant, due to high variability between individuals. Very little information is available regarding the accumulation of POPs in the flesh during a washout period; Bell et al. [16] obtained similar results. Vegetable oil-fed fish accumulated contaminants during the washout phase. Flame retardant accumulation in the flesh does not follow the path of PCBs, toxaphenes, and organochlorine pesticides (Figure 2(d)). As fish oil has a greater percentage of flame retardant levels than vegetable oils and fish were fed 100% FO diets, these data are unexpected. However, overall levels are very low (less than 1.5 ppb). In terms on POP exposure, all vegetable oil-based feeds have ability to reduce the accumulation in the flesh, in the growth phase. Even if levels were increased for PCBs, toxaphenes, and organochlorine pesticides in the washout period of this study, final concentrations were still lower than recommended levels (European Committee, 2007) [36, 37].

Digestibility of Experimental Diets

Results show that the reduction of FO content in the feed (18 FO vs 25 FO) improves ADC of dry matter and lipids (Table 7). Apparent digestibility of dry matter is in accordance with results from Windell et al. [39] for medium (ADC of 72%) and large (74%) rainbow trout fed fish meal at three different temperatures (7, 11, and 15°C). Cho et al. [40] had an ADC in a range of 70 to 77% with rainbow trout fed marine oils and different protein sources. However, Olsen et al. [41] fed 500 g Atlantic salmon with Calanus finmarchicus at 10°C seawater and obtained an ADC of 95% while our results range between 73 and 82%. Apparent digestibility was the highest for 25 CO and the lowest for 18 FO diet. Digestibility of crude protein was between 70 and 100% [39, 40, 42, 43]. Our results are not significantly different and are in accordance with these authors. Our results are within the range described by the NRC [17] where digestibility for fish oil usually varies in a range of 75 to 95%. In multiple digestibility trials with rainbow trout using feather meal, meat and bone meal, poultry by-product meal, and blood meal, Bureau et al. [44] observed digestibilities between 68 and 99% for gross energy. Cho and Kaushik [42] had an average of 76% and Cho et al. 79% [40]. In the current study, the digestibility of gross energy for experimental diets was therefore similar, ranging between 81% and 90%.

Table 7: Apparent digestibility coefficients (ADC) of experimental diets (g/100g TL)

	18 FO	18 SO	18 CO	18 COS	25 FO	25 SO	25 CO	25 COS	Pooled SEM
Dry matter	72.7^{b}	76.2^{a,b}	77.2^{a,b}	77.5^{a,b}	79.3^{a}	81.4^{a}	82.0^{a}	81.3^{a}	0.7
Crude protein	90.4	91.1	91.4	90.6	90.9	91.6	91.5	91.7	0.2
Total lipid	78.9^{c}	90.3^{a,b}	89.0^{b}	89.9^{a,b}	89.4^{b}	93.7^{a}	92.8^{a,b}	91.5^{a,b}	0.9
Gross energy	80.9^{e}	82.1^{d,e}	85.2^{c,d,e}	85.5^{b,c,d,e}	87.4^{a,b,c}	86.3^{a,b,c,d}	90.3^{a}	90.0^{a,b}	0.7

18 FO, 18 DP/DE diet with 100% fish oil; 18 SO, 18 DP/DE diet with 75% soybean oil; 18 CO, 18 DP/DE diet with 75% canola oil; 18 COS, 18 DP/DE diet with 75% blend of canola oil and Schizochytrium sp.; 25 FO, 25 DP/DE diet with 100% fish oil; 25 SO, 25 DP/DE diet with 75% soybean oil; 25 CO, 25 DP/DE diet with 75% canola oil; 25 COS, 25 DP/DE diet with 75% blend of canola oil and Schizochytrium sp.; DM, dry matter; DP/DE, digestible protein/digestible energy (lipids); TL, total lipids. Data represent the mean ± standard deviation of three replicates. Within the same line, different superscripts indicate significant differences due to the diet ($p<0.05$).

According to Caballero et al. [5], high lipid ADC for the vegetable oil-based diets is related to their high PUFA content, consequently, their low melting point. Results from Cho and Kaushik [42] support this observation with apparent digestibility of rainbow trout reared in temperatures varying from 5 to 15°C. When fed rapeseed, soybean, linseed or fish oils, ADC was high (between 80 and 95%). Conversely, when trout were fed lard and tallow, which have high melting points because of high SFA concentrations, ADC decreased compared to other lipids sources. Therefore, the melting point of lipids influences their digestibility for rainbow trout. However, this theory can not explain the variations in digestibility obtained in this study, as fish oil and vegetable oils have a similar melting point. Bureau et al. [45] reported a high ADC for tallow fed to rainbow trout, and suggested that ADC is not only a matter of the lipid's melting point but also of the "synergistic effect" of PUFA on the digestibility of SFA. Differences in results may be due to the methodology used by Cho and Kaushik [42] where experimental diets were formulated with a reference diet that was low in lipids (<5%). The addition of oils rich in PUFA would have increased the ADC of SFA, and thereby the ADC of total lipids. Overall, the balance between PUFA and SFA in the diet is crucial to obtain a high ADC. The 18 FO diet, which has the highest percentage in fish oil and total fish by-products, has the lowest digestibility.

This diet also has a high SFA/PUFA ratio (1.07) compared to vegetable oil diets (mean 0.38), confirming previous suggestions on lipid digestibility, where the SFA/PUFA ratio of the diet is crucial in the digestibility of a diet. Therefore, our results are in accordance with previous studies and Caballero et al.'s study [5] with rainbow trout reared at 12°C (ADC = 79.2 to 92.5%). In conclusion, apparent digestibility of nutrient and gross energy of the eight experimental diets, irrespective of the DP/DE ratio and the lipid source, are highly suitable for the industry, and these diets are appropriate for rainbow trout.

CONCLUSIONS

The replacement of 75% total fish oil by vegetable oils and the alteration of DP/DE ratios did not affect apparent digestibility, growth performance, and somatic parameters for trout. However, n-3/n-6 levels were negatively affected, especially for the SO groups. During the washout trial, muscle fatty acids were partially restored for the two CO and COS groups, but the SO groups still had high n-6 content. In the growth trial, POPs were lower in vegetable oil-fed trout of the 18 and 25 DP/DE groups. The washout trial increased POP percentages, but levels are below recommendation levels. Thus, replacing 75% of total fish oil in rainbow trout feed by CO and COS, combined with a 25 DP/DE ratio, seems to be the most efficient in terms of maximizing the total fish by-product replacement, while reducing POP exposure.

ACKNOWLEDGMENTS

We thank AquaNet, EWOS, Inc. (Surrey, BC), La Société de Recherche et de Développement en Aquaculture Continentale, Inc. (SORDAC) (Quebec, QC), La Société de Développement de l'Industrie Maricole, Inc. (SODIM) (Gaspé, QC), Le Réseau Aquaculture Québec (RAQ) (Rimouski, QC), Bi-Pro Marketing Ltd. (Guelph, ON), and The Canola Council of Canada (Winnipeg, MB) for financial support. Advanced BioNutrition (Columbia, MD) kindly provided the algal biomass (S-Type Gold Fat), Soya Excel (Beloeil, QC) the soybean oil, and Bunge (Montreal, QC) the canola oil. All members for the technical assistance of the Alma Aquaculture Research Station staff (AARS, Elora, ON) are acknowledged. Éric Dewailly deceased.

REFERENCES

1. R. A. Hites, J. A. Foran, D. O. Carpenter, M. C. Hamilton, B. A. Knuth, and S. J. Schwager, "Global assessment of organic contaminants in farmed salmon," Science, vol. 303, no. 5655, pp. 226–229, 2004.
2. M. H. G. Berntssen, K. Julshamn, and A. K. Lundebye, "Chemical contaminants in aquafeeds and Atlantic salmon (Salmo salar) following the use of traditional-versus alternative feed ingredients," Chemosphere, vol. 78, no. 6, pp. 637–646, 2010.
3. O. J. Nøstbakken, H. T. Hove, A. Duinker et al., "Contaminant levels in Norwegian farmed Atlantic salmon (Salmo salar) in the 13-year period from 1999 to 2011," Environment International, vol. 74, pp. 274–280, 2015.
4. E. Å. Bendiksen, C. A. Johnsen, H. J. Olsen, and M. Jobling, "Sustainable aquafeeds: progress towards reduced reliance upon marine ingredients in diets for farmed salmon (Salmo salar L.)," Aquaculture, vol. 314, no. 1–4, pp. 132–139, 2011.
5. M. J. Caballero, A. Obach, G. Rosenlund, D. Montero, M. Gisvold, and M. S. Izquiero, "Impact of different dietary lipid sources on growth, lipid digestibility, tissue fatty acid composition and histology of rainbow trout Oncorhynchus mykiss," Aquaculture, vol. 214, no. 1–4, pp. 253–271, 2002.
6. V. O. Crampton, D. A. Nanton, K. Ruohonen, P.-O. Skjervold, and A. El-Mowafi, "Demonstration of salmon farming as a net producer of fish protein and oil," Aquaculture Nutrition, vol. 16, no. 4, pp. 437–446, 2010.
7. A. E. O. Jordal, O. Lie, and B. E. Torstensen, "Complete replacement of dietary fish oil with a vegetable oil blend affect liver lipid and plasma lipoprotein levels in Atlantic salmon (Salmo salar L.)," Aquaculture Nutrition, vol. 13, no. 2, pp. 114–130, 2007.
8. P. K. Sarker, D. P. Bureau, M. Drew et al., "Sustainability issues related to feeding salmonids: a Canadian perspective," Reviews in Aquaculture, vol. 5, no. 4, pp. 199–219, 2013.
9. M. R. Shah, G. A. Lutzu, A. Alam et al., "Microalgae in aquafeeds for a sustainable aquaculture industry," Journal of Applied Phycology, vol. 30, no. 1, pp. 197–213, 2018.
10. G. M. Turchini, B. E. Torstensen, and W. K. Ng, "Fish oil replacement in finfish nutrition," Reviews in Aquaculture, vol. 1, no. 1, pp. 10–57,

2009.

11. L. Benedito-Palos, J. C. Navarro, A. Bermejo-Nogales, A. Saera-Vila, S. Kaushik, and J. Perez-Sanchez, "The time course of fish oil wash-out follows a simple dilution model in gilthead sea bream (Sparus aurata L.) fed grade levels of vegetable oils," Aquaculture, vol. 288, no. 1-2, pp. 98–105, 2009.
12. E. Fountoulaki, A. Vasilaki, R. Hurtado et al., "Fish oil substitution by vegetables oils in commercial diets for gilthead sea bream (Sparus aurata L.): effects on growth performance, flesh quality, and fillet fatty acid profile; recovery of fatty acid profiles by a fish oil finishing diet under fluctuating water temperatures," Aquaculture, vol. 289, no. 3-4, pp. 317–326, 2009.
13. J. G. Bell, F. Mcghee, P. J. Campbell, and J. R. Sargent, "Rapeseed oil as an alternative to marine fish oil in diets of post-smolt Atlantic salmon (Salmo salar): changes in flesh fatty acid composition and effectiveness of subsequent fish oil "wash out"," Aquaculture, vol. 218, no. 1–4, pp. 515–528, 2003.
14. B. E. Torstensen, L. Frøyland, R. Ørnsrud, and Ø. Lie, "Tailoring of a cardioprotective muscle fatty acid composition of Atlantic salmon (Salmo salar) fed vegetable oils," Food Chemistry, vol. 87, no. 4, pp. 567–580, 2004.
15. J. T. Trushenski, J. Boesenberg, and C. C. Kohler, "Influence of grow-out feed fatty acid composition on finishing success in Nile tilapia," North American Journal of Aquaculture, vol. 71, no. 3, pp. 242–251, 2009.
16. J. G. Bell, F. Mcghee, J. R. Dick, and D. R. Tocher, "Dioxin and dioxin-like polychlorinated biphenyls (PCBs) in Scottish farmed salmon (Salmo salar): effects of replacement of dietary marine fish oil with vegetable oils," Aquaculture, vol. 243, no. 1–4, pp. 305–314, 2005.
17. National Research Council (NRC), Nutrient Requirements of Fish, National Academy Press, Washington, DC, USA, 1993.
18. P. A. Azevedo, J. Van Milgen, S. Leeson, and D. P. Bureau, "Comparing efficiency of metabolizable energy utilization by rainbow trout (Oncorhynchus mykiss) and Atlantic salmon (Salmo salar) using factorial and multivariate approaches," Journal of Animal Science, vol. 83, no. 4, pp. 842–851, 2005.
19. AOAC, AOAC Official Methods of Analysis, Association of Official

Analytical Chemists, Method number 927.05, 930.30, Rockville, MD, USA, 15th edition, 1990.

20. AOAC, AOAC Official Method 996.06. Fat (Total, Saturated, and Unsaturated) in Foods: Hydrolytic Extraction Gas Chromatographic Method, AOAC Official Methods of Analysis, Rockville, MD, USA, 2000.
21. C. Y. Cho, S. J. Slinger, and H. S. Bayley, "Bioenergetics of salmonid fishes: energy intake, expenditure and productivity," Comparative Biochemistry and Physiology Part B: Comparative Biochemistry, vol. 73, no. 1, pp. 25–41, 1982.
22. C. Y. Cho and S. J. Slinger, "Apparent digestibility measurement in feedstuffs for rainbow trout," in Proceedings of World Symposium on Finfish Nutrition and Fishfeed Technology, vol. 2, pp. 239–247, Hamburg, Germany, June 1979.
23. R. J. Henderson, "Fatty acid metabolism in freshwater fish with particular reference to polyunsaturated fatty acids," Archive for Animal Nutrition, vol. 49, no. 1, pp. 5–22, 1996.
24. X. X. Wang, Y. J. Li, C. L. Hou, Y. Gao, and Y. Z. Wang, "Influence of different dietary lipid sources on the growth, tissue fatty acid composition, histological changes and peroxisome proliferator-activated receptor γ gene expression in large yellow croaker (Pseudosciaena crocea R.)," Aquaculture Research, vol. 43, no. 2, pp. 281–291, 2011.
25. B. Ruyter, C. Røsjøa, B. Grisdale-Helland, G. Rosenlund, A. Obach, and M. S. Thomassena, "Influence of temperature and high dietary linoleic acid content on esterification, elongation, and desaturation of PUFA in Atlantic salmon hepatocytes," Lipids, vol. 38, no. 8, pp. 833–840, 2003.
26. J. G. Bell, I. Ashton, C. J. Secombes, B. R. Weitzel, and J. R. Dick, "Dietary lipid affects phospholipid fatty acid compositions, eicosanoid production and immune function in Atlantic salmon (Salmo salar)," Prostaglandins, Leukotrienes and Essential Fatty Acids, vol. 54, no. 3, pp. 173–182, 1996.
27. B. D. Glencross, "Exploring the nutritional demand for essential fatty acids by aquaculture species," Reviews in Aquaculture, vol. 1, no. 2, pp. 71–124, 2009.
28. D. H. S. Greene and D. P. Selivonchick, "Effects of dietary vegetable,

animal and marine lipids on muscle lipid and hematology of rainbow trout (Oncorhynchus mykiss)," Aquaculture, vol. 89, no. 2, pp. 165–182, 1990.

29. C. Moya-Falcón, E. Hvattum, T. N. Tran, M. S. Thomassen, M. S. Skorve, and B. Ruyter, "Phospholipid molecular species, β-oxidation, desaturation and elongation of fatty acids in Atlantic salmon hepatocytes: effects of temperature and 3-thia fatty acids," Comparative Biochemistry and Physiology Part B: Biochemistry and Molecular Biology, vol. 145, no. 1, pp. 68–80, 2006.
30. J. R. Sargent, D. R. Tocher, and J. G. Bell, "Fish nutrition," in The Lipids, J. E. Halver and R. W. Hard, Eds., Academic Press, Cambridge, MA, USA, 2002.
31. B. M. Anderson and D. W. L. Ma, "Are all n-3 polyunsaturated fatty acids created equal?" Lipids in Health and Disease, vol. 8, no. 1, p. 33, 2009.
32. A. P. Simopoulos, "Essential fatty acids in health and chronic disease," American Journal of Clinical Nutrition, vol. 70, no. 3, pp. 560S–590S, 1999.
33. E. L. Charnov, T. F. Turner, and K. O. Winemiller, "Reproductive constraints and the evolution of life histories with indeterminate growth," Proceedings of the National Academy of Sciences, vol. 98, no. 16, pp. 9460–9464, 2001.
34. T. Day and P. D. Taylor, "Von Bertalanffy's growth equation should not be used to model age and size at maturity," American Naturalist, vol. 149, no. 2, pp. 381–393, 1997.
35. A. Dumas, J. France, and D. P. Bureau, "Evidence of three growth stanzas in rainbow trout (Oncorhynchus mykiss) across life stages and adaptation of the thermal-unit growth coefficient," Aquaculture, vol. 267, no. 1–4, pp. 139–146, 2007.
36. Health Canada, 2007, http://www.cfsan.fda.gov/□acrobat/haccpc09.pdf.
37. Food and Drug Administration (FDA), 2017, http://www.fda.gov/ora/compliance_ref/cpg/cpgfod/cpg575-100.html.
38. E. N. Friesen, M. G. Ikonomou, D. A. Higgs, K. P. Ang, and C. Dubetz, "Use of terrestrial based lipids in aquaculture feeds and the effects on flesh organohalogen and fatty acid concentrations in farmed Atlantic salmon," Environmental Science & Technology, vol. 42, no. 10, pp.

3519–3523, 2008.

39. J. T. Windell, J. W. Foltz, and J. A. Sarokon, "Effect of fish size, temperature, and amount fed on nutrient digestibility of a pelleted diet by rainbow trout, Salmo gairdneri," Transactions of the American Fisheries Society, vol. 107, no. 4, pp. 613–616, 1978.
40. C. Y. Cho, H. S. Bayley, and S. J. Slinger, "Partial replacement of herring meal with soybean meal and other changes in a diet for rainbow trout (Salmo gairdneri)," Journal of the Fisheries Research Board of Canada, vol. 31, no. 9, pp. 1523–1528, 1974.
41. R. E. Olsen, R. J. Henderson, J. Sountama et al., "Atlantic salmon, Salmo salar, utilizes wax ester-rich oil from Calanus finmarchicus effectively," Aquaculture, vol. 240, no. 1–4, pp. 433–449, 2004.
42. C. Y. Cho and S. J. Kaushik, "Nutritional energetics in fish: energy and protein utilization in rainbow trout (Salmo gairdneri). Aspects of food production, consumption and energy values," in World Review of Nutrition and Dietetics, G. H. Bourne, Ed., vol. 61, pp. 132–172, Karger, Basel, Switzerland, 1990.
43. T. G. Gaylord, F. T. Barrows, and S. D. Rawles, "Apparent digestibility of gross nutrients from feedstuffs in extruded feeds for rainbow trout, Oncorhynchus mykiss," Journal of the World Aquaculture Society, vol. 39, no. 6, pp. 827–834, 2008.
44. D. P. Bureau, A. M. Harris, and C. Y. Cho, "Apparent digestibility of rendered animal protein ingredients for rainbow trout (Oncorhynchus mykiss)," Aquaculture, vol. 180, no. 3-4, pp. 345–358, 1999.
45. D. P. Bureau, K. Hua, and A. M. Harris, "The effect of dietary lipid and long-chain n-3 PUFA levels on growth, energy utilisation, carcass quality, and immune function of rainbow trout, Oncorhynchus mykiss," Journal of the World Aquaculture Society, vol. 39, no. 1, pp. 1–21, 2008.
46. Environnement Canada, 2017, http://www.msc.ec.gc.ca.
47. S. Koshio, R. G. Ackman, and S. P. Lall, "Effects of oxidized herring and canola oils in diets on growth, survival, and flavor of Atlantic salmon, Salmo salar," Journal of Agricultural and Food Chemistry, vol. 42, no. 5, pp. 1164–1169, 1994.
48. M. S. Thomassen and C. Røsjø, "Different fats in feed for salmon: influence on sensory parameters, growth rate and fatty acids in muscle and heart," Aquaculture, vol. 79, no. 1–4, pp. 129–135, 1989.

CHAPTER 6

IS LIPID PROFILE ASSOCIATED WITH BONE MINERAL DENSITY AND BONE FORMATION IN SUBJECTS WITH SPINAL CORD INJURY?

Hadis Sabour[1] , Abbas Norouzi Javidan[1] , Sahar Latifi[1] , Mohammad Reza Hadian[1] , Seyed-Hassan Emami Razavi[1] , Farzad Shidfar[2] , Mohammad Reza Vafa[2] , and Hamidreza Aghaei Meybodi[3]

[1]Brain and Spinal Injury Research Center (BASIR), Tehran University of Medical Sciences, Tehran, Iran
[2]Nutrition Department, Iran University of Medical Sciences, Tehran, Iran
[3]Endocrinology and Metabolism Research Institute (EMRI), Tehran University of Medical Sciences, Tehran, Iran

ABSTRACT

Purpose. The association between serum lipids and bone mineral density (BMD) has been investigated previously but, up to now, these relationships have not yet been described in spinal cord injury (SCI). We tried to assess the correlation between serum triglyceride (TG), total cholesterol (TC),

Citation: Hadis Sabour,Abbas Norouzi Javidan,Sahar Latifi, Mohammad Reza Hadian, Seyed-Hassan Emami Razavi, Farzad Shidfar, Mohammad Reza Vafa, and Hamidreza Aghaei Meybodi. Is Lipid Profile Associated with Bone Mineral Density and Bone Formation in Subjects with Spinal Cord Injury? Journal of Osteoporosis. Volume 2014, Article ID 695014, 6 pages. http://dx.doi.org/10.1155/2014/695014.

high-density lipoprotein (HDL), and low-density lipoprotein (LDL) and BMD in male subjects with SCI. Methods. Dual-energy X-ray absorptiometry (DXA) was used to assess BMD in femoral neck, trochanter, intertrochanteric zone, and lumbar vertebras. Blood samples were taken to measure serums lipids and bone biomarkers including osteocalcin, cross-linked type I collagen (CTX), and bone alkaline phosphatase (BALP). Partial correlation analysis was used to evaluate the relationships between mentioned measurements after adjustment for weight and age. Results. We found a positive correlation between HDL and femoral neck BMD (P: 0.004, $r = 0.33$). HDL was negatively correlated with osteocalcin (P: 0.017, $r = -0.31$) which was not in consistency with its relationship with BMD. TC and LDL were not related to CTX, BALP and BMD. Conclusion. This study does not support a strong association between serum lipids and BMD in subjects with SCI. Moreover it seems that positive association between HDL and BMD is not mediated through increased bone formation.

INTRODUCTION

Existence of a probable relationship between lipid profile and bone mineral density (BMD) was initially derived from reported results on statins' positive effect on BMD [1, 2]. Various studies have supported the promising influence of statins (the HMG co-A reductase inhibitors) on reduction of fractures along with increased BMD [3, 4]. The assumption of relationship between lipids and BMD was tested in two cohort studies and showed a significant positive association between BMD and low density lipoprotein (LDL) and total triglyceride (TG) while it was negatively correlated with high density lipoprotein (HDL) [5]. However some other literatures showed no relationship between lipid profile and BMD [6]. Most of these studies were on healthy individuals. Spinal cord injury (SCI) is a severe medical condition which restricts physical activities tremendously [7]. Along with immobility, it is associated with some changes in lipid profile [8–10]. In this regard, de Groot et al. demonstrated decreased TC and elevated HDL during and one year after rehabilitation; however these patients may also experience increased levels of TC and LDL after discharge [8]. The study of de Groot also revealed the effect of age in subjects with SCI since higher TG, LDL, and HDL were observed in older patients. Wong et al. [9] reported the decreased level of HDL in subjects with SCI and their study also illustrated the effect of injury level since higher TG was detected among patients with injury at lumbar level. Moreover, osteoporosis occurs among these patients due to mechanical unloading [11–14] along with alterations in bone metabolism [15]. To our knowledge, the effect of serum lipids on BMD in subjects with SCI is still unknown and mostly the assessed population was postmenopausal women [16–18] which is a known population with higher risk of osteoporosis. In this study, we investigated only male subjects to omit the bias effect of menopause and other osteoporosis gender-related risk factors. Subjects with SCI also represent a sensitive population susceptible to osteoporosis and evaluation of the correlation between lipids and BMD in these patients is the main purpose of this study.

Changes in bone mass are associated with alterations in bone biomarkers. Lipids may affect BMD through their influence on bone biomarkers. In this field there are some specific known biomarkers including cross-linked type I collagen (CTX), osteocalcin, and bone alkaline phosphatase (BALP). CTX represents bone resorption and its increased level has been shown in SCI [19]. Osteocalcin and BALP are associated with bone formation. Our previous study showed that osteocalcin is negatively related to the BMD of femoral intertrochanteric zone [15]. We also demonstrated the positive association between CTX, BALP, and osteocalcin which shows the coincidental occurrence of osteoblastic and osteoclastic activities among subjects with SCI [15]. In the present study, we tried to assess correlations between bone biomarkers and lipid profile. Up to now, the association between BMD and lipid profile in subjects with SCI has not yet been reported and here in this study we tried to evaluate this relationship.

MATERIALS AND METHODS

Study Design

This study was a cross-sectional investigation which was performed in Brain and Spinal Injury Research Center (BASIR). Data was collected from November 2010 to 2011. All patients received adequate information about the study and written consents were obtained before enrollment. The protocol was approved by the ethics committee at Tehran University of Medical Sciences.

Participants

Referred patients with SCI were invited to participate in this study. Inclusion criteria were male subjects, SCI with traumatic etiologies, and time since injury of more than 1 year. Exclusion criteria were pregnancy, lactation, amputation, recent fracture, and nontraumatic SCI etiology. Patients with history of diabetes, cancer, endocrinology disease, acute infection, coronary artery diseases, use of special medications such as glucocorticoid, hormones, thyroid hormones, anticonvulsive drugs, heparin, aluminum containing antacids, lithium, omega-3 fatty acids, or other nutrients supplements, and smoking or alcohol consumption were also excluded.

Anthropometric Measurements

Patients' demographic characteristics including age and postinjury duration were obtained during face-to-face interviews. Body weight was measured using a digital wheelchair scale, body height was obtained measuring the supine length, and body mass index (BMI) was calculated as body weight (in kilograms) divided by height (in meters) squared.

Bone Mineral Density Measurements

Dual-energy X-ray absorptiometry (DXA) was used to assess BMD in three bone sites of femur (neck, intertrochanteric zone, and trochanter) and spinal lumbar vertebras. Total hip BMD was measured as well. Calibration of bone densitometer Lunar DPXMD device (Lunar Corporation, Madison, WI, USA) was performed weekly by using appropriate phantoms. The precision error (PE) is usually expressed as the coefficient of variation (CV), which is the ratio of the SD to the mean of the measurements [20]. The precision error for bone mineral density measurements was 2 to 3% in the femoral and 1 to 1.5% in the lumbar regions. All scans were performed according to the manufacturer's guidelines. In patients with spinal implant the involved lumbar vertebras were excluded and the mean bone density of noninvolved vertebras was entered into the analysis.

Laboratory Measurements

Blood samples were taken under antiseptic conditions from antecubital vein. Blood samples were collected and centrifuged at 3000 rpm for 10 minutes at 4°C. Single session analysis was used to reduce interassay variation in serum samples. Samples were sent to the Endocrinology and Metabolism Research Center (EMRC) laboratory for analysis and were frozen immediately. Serums HDL, LDL, total cholesterol (TC), and total triglyceride (TG) were measured. Circulating levels of osteocalcin and C-telopeptide cross-linked type 1 collagen (CTX) were quantified by electrochemiluminescence immunoassay with detection limits of 0.50 ng/mL for osteocalcin and 0.07 ng/mL for CTX. BALP was measured by bone-specific alkaline phosphatase (BALP) ELISA kit with detection range of 1.25 ng/mL–80 ng/mL. All measurements were performed in Endocrinology and Metabolism Research Center (EMRC).

Statistical Analysis

All statistical analysis was performed by SPSS version 21 (IBM Corporation). The relationship between lipid profile indexes and anthropometric variables (age, weight, BMI, and time since injury) was assessed by simple linear regression. The association between BMD in different bone sites and lipids was evaluated using partial correlation with adjustment for weight, height, BMI, and age. P value < 0.05 was considered statistically significant.

RESULTS

Total of 85 subjects with mean age of 51.80 ± 13.44 entered this investigation. Basic characteristics of these patients are summarized in Table 1. Eighteen subjects (21.2%) had injury at cervical level, 50 (58.8%) at thoracic level, and 17 (20%) at lumbar level. Mean TG was slightly higher among subjects with injury at lumbar level (170.29 ± 85 in lumbar group and 151.54±58.97 and 138.55±53 in thoracic and cervical groups, resp.,) but this difference was not statistically significant (*P*: 0.34). TC, HDL, and LDL also showed no relationship with injury level (*P*: 0.74, 0.68, and 0.71, resp.).

Table 1: Baseline characteristics of study cases and differences of basic features among male subjects with spinal cord injury (: 85)

Category		Mean (standard deviation)
Age (year)		51.80 (13.44)
Weight (kg)		70.49 (14.36)
Height (cm)		172.34 (6.92)
Body mass index (kg/m^2)		23.62 (4.05)
Time since injury (year)		9.15 (6.43)
Triglyceride (mg/dL)		152.54 (64.06)
Total cholesterol (mg/dL)		175.69 (34.38)
High density lipoprotein (mg/dL)		41.31 (8.94)
Low density lipoprotein (mg/dL)		100.49 (24.97)

Bone mineral density of femoral neck	T-score	−2.16 (0.95)
	Z-score	−1.74 (0.96)
Bone mineral density of femoral intertrochanteric zone	T-score	−2.29 (0.94)
	Z-score	−2.21 (0.95)
Bone mineral density of femoral trochanter	T-score	−2.11 (0.73)
	Z-score	−1.92 (0.74)
Bone mineral density of total hip	T-score	−2.36 (0.83)
	Z-score	−2.22 (0.86)
Bone mineral density of spinal lumbar vertebras	T-score	−0.43 (1.81)
	Z-score	0.53 (1.59)

Mean triglyceride was 152.53±64.06 and mean of TC was 175.69±34.38. Means of HDL and LDL were 41.31±8.94 and 100.49 ± 24.97, respectively

The primary analysis of confounders effects including age, weight, height, and BMI on serum lipids showed that there is a significant negative association between weight and HDL (P: 0.002, $r = -0.32$) and also between age and TC as well as LDL ($P < 0.0001$, $r = -0.38$ and $P < 0.0001$, $r = -0.36$, resp.). Weight was positively related to TG (P: 0.003, $r = 0.31$). BMI was correlated with LDL (P: 0.024, $r = 0.24$), TG (P: 0.007, $r = 0.29$), and HDL (P: 0.009, $r = -0.28$). The existence of these statistically significant relationships defines these mentioned factors as confounder [21], so, in the next step of analysis, the relationship between serum lipids and BMD and bone biomarkers was evaluated after adjustment for mentioned factors.

After adjustment for age, weight, height, and BMI, the correlation between serum lipids and BMD was assessed. The effect of lipids on bone biomarkers was evaluated as well. We found a positive correlation only between HDL and femoral neck BMD (Table 2). Furthermore, we did not find any significant effect of serum lipids on CTX and BALP (Table 2) while HDL was negatively correlated with osteocalcin. The results are interesting as the positive effect of HDL on femoral neck BMD is not consistent with negative association between osteocalcin and HDL because mostly we consider osteocalcin as a bone formation indicator. These results show that serum lipids and especially HDL may affect long bone BMD in patients with SCI through mechanisms which are not merely mediated by pathways in which bone biomarkers' alterations are involved.

Table 2: The relationship between serum lipids and bone mineral density in male subjects with spinal cord injury

Category		Male subjects with spinal cord injury (n: 85)			
		TG	TC	HDL	LDL
Femoral neck BMD	T-score	0.17	0.84	0.004** (r = 0.33)	0.42
	Z-score	0.18	0.81	0.004** (r = 0.35)	0.45
Femoral intertrochanteric zone BMD	T-score	0.28	0.49	0.08	0.90
	Z-score	0.29	0.45	0.07	0.86
Femoral trochanter BMD	T-score	0.88	0.73	0.97	0.95
	Z-score	0.87	0.71	0.90	0.94
Total hip BMD	T-score	0.36	0.42	0.09	0.85
	Z-score	0.40	0.41	0.09	0.85
Spinal lumbar vertebras' BMD	T-score	0.56	0.52	0.78	0.27
	Z-score	0.73	0.74	0.31	0.95
Osteocalcin		0.99	0.22	0.017* (r = −0.31)	0.91
Cross-linked type I collagen		0.56	0.79	0.64	0.97
Bone alkaline phosphatase		0.76	0.65	0.92	0.40

BMD: bone mineral density, HDL: high density lipoprotein, LDL: low density lipoprotein, TC: total cholesterol, and TG: total triglyceride.

*Significance at level of P<0.05.

**Significance at level of P<0.01.

DISCUSSION

The probable association between serum lipids and BMD has been proposed when the positive effect of lipid reducing medications such as statins in increasing BMD was observed in various studies [22, 23]. However, while Wada et al. [24] detected no association between statin use and lumbar BMD, other literatures have shown that reduced fracture risk is greater than statin-induced increased BMD [25]. These correlations seem to be more evident in patients with increased risk of osteoporosis rather than healthy able-bodies [16, 17, 26]. The hypothesis of the association between atherogenic lipid profiles and BMD has been tested in several studies and the controversial results have been published up to now. Some investigations supported the negative association between atherogenic lipid profile and BMD [18], some found no relationship [6, 27], and some studies reported positive correlation [5, 17]. Solomon et al. [6] investigated a large number of able-bodies in a national survey and found no association between serum lipids and BMD. This study is noticeable by considering its sample size. However, these relationships may be different in individuals with background diseases or medical conditions that are known to be associated with osteoporosis. Here, in male subjects with SCI, we found that femoral neck BMD was positively related to HDL. Previously Dennison et al. [28]

showed a negative correlation between HDL and femoral BMD in healthy individuals which contradicts with our results in subjects with SCI. The results of de Groot et al. [8] showed that, after SCI, an elevated level of TC and LDL can be observed which can be reduced during rehabilitation. This unfavorable hyperlipidemia that occurs after SCI may play a part in the controversial results between our study and Dennison's investigation on healthy subjects. Wong et al. [9] illustrated the effect of injury level on lipid profile and their study showed higher level of TC among patients with injury at lumbar level. We found no significant relationship between injury level and TC which contradicts with Wong's study. Many various factors such as ethnicity, diets, and life-style affect lipid profile and may result in such contradictions between our results in Iranian patients and Wong's study in Chinese population. It has been shown that lipid profile in subjects with SCI goes under some alterations after SCI which is mostly toward reduction of HDL in these patients [29–31]. The negative correlation between HDL and BMD which was reported in mentioned studies does not lead to increased BMD in subjects with SCI because bone loss in these patients is a consequence of various factors including mechanical unloading and changes in bone metabolism. As a result we do not observe such a negative association between HDL and BMD in subjects with SCI. The controversial results in different study populations suggest that lipids' relationship with BMD is manipulated by many various factors and, apart from background medical condition, factors like lifestyle, physical activity, and amount of fat mass should be considered.

Most studies found no association between TC and BMD in various populations [27, 32, 33] which is in agreement with our results in SCI. However, while Parhami et al. [34] illustrated that baseline level of cholesterol synthesis is necessary for the osteoblastic differentiation, it seems that total cholesterol does not directly affect BMD. Here, we report the same findings in subjects with SCI.

CTX which represents bone resorption and whose negative association with BMD has been frequently demonstrated [35, 36] does not seem to be related to serum lipids. BALP which is a factor reflecting bone formation did not show any relationship with lipids. However, osteocalcin which is a specific marker illustrating bone formation and osteoblastic activity [37] was negatively associated with HDL in men. These results show that positive association between HDL and BMD in male subjects with SCI is not mediated through pathways which involve increased osteocalcin. In fact this result suggests that HDL positive relationship with BMD does not affect bone formation markers, which proposes that this effect is not mediated through increased bone formation.

LDL was also not related to BMD in subjects with SCI, which is in agreement with Dennison's report in general population [28]. Many studies do not support strong association between lipids' concentrations and bone mass measurements [38]. In this study, we report also a weak relationship between lipid levels and BMD in SCI population. Only HDL was positively associated with femoral neck BMD. LDL and TC levels revealed no influence on BMD.

CONCLUSION

This study does not support a strong association between serum lipids and BMD in patients with SCI. LDL and TC had no relationship with BMD in male subjects with SCI while HDL was positively associated with femoral neck BMD. This positive correlation between HDL and BMD was not in line with bone biomarkers' alterations as osteocalcin, which is a bone formation marker, was negatively associated with HDL and no other relationship with CTX and BALP was observed. These results suggest that the relationship between serum lipids and BMD in subjects with SCI is weak. This study shows a positive correlation between HDL and femoral neck bone mineral density (BMD) and a negative association between HDL and osteocalcin in male subjects with spinal cord injury (SCI). TC and LDL were not related to BMD and bone biomarkers in SCI.

AUTHORS' CONTRIBUTION

Hadis Sabour contributed in study design, Abbas Norouzi Javidan helped in recruiting the patients and collecting the data, Sahar Latifi contributed in statistical analysis and writing the paper, Mohammad Reza Hadian contributed in editing the paper, Seyed-Hassan Emami Razavi obtained the ethical approve, Farzad Shidfar helped in study design, and Mohammad Reza Vafa contributed in paper critical editing in intellectual contents. Hamidreza Aghaei Meybodi contributed in data interpretation and study designing.

ACKNOWLEDGMENTS

This investigation was financially supported by a grant from Tehran University of Medical Sciences (Grant no. 7538/5/2008). This study was performed in Brain and Spinal Injury Research Center (BASIR) and the authors would like to thank all of the staff that helped them in data collection and for their cooperation in performing this study.

REFERENCES

1. P. S. Wang, D. H. Solomon, H. Mogun, and J. Avorn, "HMG-CoA reductase inhibitors and the risk of hip fractures in elderly patients," The Journal of the American Medical Association, vol. 283, no. 24, pp. 3211–3216, 2000.
2. G. Mundy, R. Garrett, S. Harris et al., "Stimulation of bone formation in vitro and in rodents by statins," Science, vol. 286, no. 5446, pp. 1946–1949, 1999.
3. C. R. Meier, R. G. Schlienger, M. E. Kraenzlin, B. Schlegel, and H. Jick, "HMG-CoA reductase inhibitors and the risk of fractures," The Journal of the American Medical Association, vol. 283, no. 24, pp. 3205–3210, 2000.
4. M. W. C. J. Schoofs, M. C. J. M. Sturkenboom, M. van der Klift, A. Hofman, H. A. P. Pols, and B. H. C. Stricker, "HMG-CoA reductase inhibitors and the risk of vertebral fracture," Journal of Bone and Mineral Research, vol. 19, no. 9, pp. 1525–1530, 2004.
5. S. Adami, V. Braga, M. Zamboni et al., "Relationship between lipids and bone mass in 2 cohorts of healthy women and men," Calcified Tissue International, vol. 74, no. 2, pp. 136–142, 2004.
6. D. H. Solomon, J. Avorn, C. F. Canning, and P. S. Wang, "Lipid levels and bone mineral density," The American Journal of Medicine, vol. 118, no. 12, pp. 1414–e5, 2005.
7. M. Midha, J. K. Schmitt, and M. Sclater, "Exercise effect with the wheelchair aerobic fitness trainer on conditioning and metabolic function in disabled persons: a pilot study," Archives of Physical Medicine and Rehabilitation, vol. 80, no. 3, pp. 258–261, 1999.
8. S. de Groot, A. J. Dallmeijer, M. W. Post, E. L. Angenot, R. J. van den Berg-Emons, and L. H. van der Woude, "Prospective analysis of lipid profiles in persons with a spinal cord injury during and 1 year after inpatient rehabilitation," Archives of Physical Medicine and Rehabilitation, vol. 89, no. 3, pp. 531–537, 2008.
9. P. Y. Wong, Y. C. Wang, N. K. Chu, F. T. Tang, and M. K. Wong, "Body weight, serum uric acid and lipid profile one year after spinal cord injury," Chang Gung medical journal, vol. 24, no. 9, pp. 569–575, 2001.·
10. W. A. Bauman, R. H. Adkins, A. M. Spungen, P. Maloney, R. Gambino, and R. L. Waters, "Ethnicity effect on the serum lipid profile in

persons with spinal cord injury," Archives of Physical Medicine and Rehabilitation, vol. 79, no. 2, pp. 176–180, 1998.

11. R. A. Battaglino, A. A. Lazzari, E. Garshick, and L. R. Morse, "Spinal cord injury-induced osteoporosis: pathogenesis and emerging therapies," Current Osteoporosis Reports, vol. 10, no. 4, pp. 278–285, 2012.·
12. C. Fattal, D. Mariano-Goulart, E. Thomas, H. Rouays-Mabit, C. Verollet, and L. Maimoun, "Osteoporosis in persons with spinal cord injury: the need for a targeted therapeutic education," Archives of Physical Medicine and Rehabilitation, vol. 92, no. 1, pp. 59–67, 2011.
13. C. Charmetant, V. Phaner, A. Condemine, and P. Calmels, "Diagnosis and treatment of osteoporosis in spinal cord injury patients: a literature review," Annals of Physical and Rehabilitation Medicine, vol. 53, no. 10, pp. 655–668, 2010.
14. S. Jiang, L. Jiang, and L. Dai, "Mechanisms of osteoporosis in spinal cord injury," Clinical Endocrinology, vol. 65, no. 5, pp. 555–565, 2006.
15. H. Sabour, A. Norouzi Javidan, S. Latifi et al., "Bone biomarkers in patients with chronic traumatic spinal cord injury," The Spine Journal, vol. 14, no. 7, pp. 1132–1138, 2014.
16. F. Sivas, E. Alemdaro□lu, E. Elverici, T. Kulu□, and K. Özoran, "Serum lipid profile: its relationship with osteoporotic vertebrae fractures and bone mineral density in Turkish postmenopausal women," Rheumatology International, vol. 29, no. 8, pp. 885–890, 2009.
17. R. A. Brownbill and J. Z. Ilich, "Lipid profile and bone paradox: higher serum lipids are associated with higher bone mineral density in postmenopausal women," Journal of Women's Health, vol. 15, no. 3, pp. 261–270, 2006.
18. P. Orozco, "Atherogenic lipid profile and elevated lipoprotein (a) are associated with lower bone mineral density in early postmenopausal overweight women," European Journal of Epidemiology, vol. 19, no. 12, pp. 1105–1112, 2004.
19. S. Bağış, G. Şahin, C. Aybay, and A. Karagöz, "Bone metabolism in patients with spinal cord injury," Turkish Journal of Rheumatology, vol. 17, pp. 168–173, 2002.
20. A. El Maghraoui and C. Roux, "DXA scanning in clinical practice," QJM, vol. 101, no. 8, pp. 605–617, 2008.

21. K. J. Rothman and S. Greenland, Modern Epidemiology, Lippincott Williams and Wilkins, Philadelphia, Pa, USA, 2nd edition, 1998.
22. J. L. Hernández, J. M. Olmos, G. Romaña et al., "Bone mineral density in statin users: a population-based analysis from a Spanish cohort," Journal of Bone and Mineral Metabolism, vol. 32, no. 2, pp. 184–191, 2014.
23. S. Chuengsamarn, S. Rattanamongkoulgul, S. Suwanwalaikorn, S. Wattanasirichaigoon, and L. Kaufman, "Effects of statins vs. non-statin lipid-lowering therapy on bone formation and bone mineral density biomarkers in patients with hyperlipidemia," Bone, vol. 46, no. 4, pp. 1011–1015, 2010.
24. Y. Wada, Y. Nakamura, and H. Koshiyama, "Lack of positive correlation between statin use and bone mineral density in Japanese subjects with type 2 diabetes," Archives of Internal Medicine, vol. 160, no. 18, p. 2865, 2000.
25. J. A. Pasco, M. A. Kotowicz, M. J. Henry, K. M. Sanders, and G. C. Nicholson, "Statin use, bone mineral density, and fracture risk: Geelong Osteoporosis Study," Archives of Internal Medicine, vol. 162, no. 5, pp. 537–540, 2002.
26. G. V. R. Prasad, R. Chiu, M. M. Nash, and J. S. Zaltzman, "Statin use and bone mineral density in renal transplant recipients," American Journal of Transplantation, vol. 3, no. 10, pp. 1320–1321, 2003.
27. E. J. Samelson, L. A. Cupples, M. T. Hannan et al., "Long-term effects of serum cholesterol on bone mineral density in women and men: the Framingham Osteoporosis Study," Bone, vol. 34, no. 3, pp. 557–561, 2004.
28. E. M. Dennison, H. E. Syddall, A. Aihie Sayer, H. J. Martin, and C. Cooper, "Lipid profile, obesity and bone mineral density: the Hertfordshire Cohort Study," QJM, vol. 100, no. 5, pp. 297–303, 2007.
29. W. A. Bauman, A. M. Spungen, Y. G. Zhong, J. L. Rothstein, C. Petry, and S. K. Gordon, "Depressed serum high density lipoprotein cholesterol levels in veterans with spinal cord injury," Paraplegia, vol. 30, no. 10, pp. 697–703, 1992.
30. G. Brenes, S. Dearwater, R. Shapera, R. E. LaPorte, and E. Collins, "High density lipoprotein cholesterol concentrations in physically active and sedentary spinal cord injured patients," Archives of Physical Medicine and Rehabilitation, vol. 67, no. 7, pp. 445–450, 1986.

31. H. Krum, L. G. Howes, D. J. Brown et al., "Risk factors for cardiovascular disease in chronic spinal cord injury patients," Paraplegia, vol. 30, no. 6, pp. 381–388, 1992.
32. L. B. Tankó, Y. Z. Bagger, S. B. Nielsen, and C. Christiansen, "Does serum cholesterol contribute to vertebral bone loss in postmenopausal women?" Bone, vol. 32, no. 1, pp. 8–14, 2003.
33. L. Y. Wu, T. C. Yang, S. W. Kuo et al., "Correlation between bone mineral density and plasma lipids in Taiwan," Endocrine Research, vol. 29, no. 3, pp. 317–325, 2003.
34. F. Parhami, N. Mody, N. Gharavi, A. J. Ballard, Y. Tintut, and L. L. Demer, "Role of the cholesterol biosynthetic pathway in osteoblastic differentiation of marrow stromal cells," Journal of Bone and Mineral Research, vol. 17, no. 11, pp. 1997–2003, 2002.
35. P. Garnero, M. Ferreras, M. A. Karsdal et al., "The type I collagen fragments ICTP and CTX reveal distinct enzymatic pathways of bone collagen degradation," Journal of Bone and Mineral Research, vol. 18, no. 5, pp. 859–867, 2003.
36. H. N. Rosen, A. C. Moses, J. Garber et al., "Serum CTX: a new marker of bone resorption that shows treatment effect more often than other markers because of low coefficient of variability and large changes with bisphosphonate therapy," Calcified Tissue International, vol. 66, no. 2, pp. 100–103, 2000.
37. J. B. Lian and C. M. Gundberg, "Osteocalcin. Biochemical considerations and clinical applications," Clinical Orthopaedics and Related Research, no. 226, pp. 267–291, 1988.
38. P. Pliatsika, A. Antoniou, A. Alexandrou et al., "Serum lipid levels and bone mineral density in Greek postmenopausal women," Gynecological Endocrinology, vol. 28, no. 8, pp. 655–660, 2012.

CHAPTER 7

LIPID DROPLETS AND MYCOBACTERIUM LEPRAE INFECTION

Ayssar A. Elamin[1] , Matthias Stehr[1] , and Mahavir Singh[1,2]

[1]Department of Research and Development, LIONEX Diagnostics and Therapeutics GmbH, 38126 Braunschweig, Germany
[2]Department of Genome Analytics, Helmholtz Centre for Infection Research, Inhoffenstrasse 7, 38124 Braunschweig, Germany

ABSTRACT

Leprosy is a chronic infectious disease and is a major source of morbidity in developing countries. Leprosy is caused by the obligate intracellular bacterium Mycobacterium leprae, which infects as primary target Schwann cells. Lepromatous leprosy exhibits multiple lesions of the skin, eyes, nerves, and lymph nodes. The sites of infection are characterized by the presence of foamy macrophages, fully packed with lipid droplets (LDs), which are in-

Citation: Ayssar A. Elamin, Matthias Stehr, and Mahavir Singh. Lipid Droplets and Mycobacterium leprae Infection, Hindawi Publishing Corporation Journal of Pathogens Volume 2012, Article ID 361374, 10 pages doi:10.1155/2012/361374

duced by M. leprae. In the last years, it has become evident that M. tuberculosis imports lipids from foamy macrophages and is dependent on fatty acids for growth in infected macrophages. M. leprae seems to have similar mechanisms for scavenging lipids from the host. But due to the inability to culture M. leprae on laboratory media, research progresses only slowly. However, in the last years, substantial progress has been made in the field of lipid metabolism in M. leprae. Herein, we will present and summarize the lipid droplets formation and the metabolism of lipids during M. leprae infection.

INTRODUCTION

Leprosy, a major source of morbidity in developing countries, is a chronic infectious disease caused by the obligate intracellular bacterium Mycobacterium leprae [1, 2]. According to the system of classification of Ridley and Jopling (1966), leprosy patients show two major manifestations of the disease, designated as lepromatous leprosy (LL) and tuberculoid leprosy (TT) [1]. TT is observed in patients with good T-cell mediated (Th1) immunity and is characterized by granuloma formation and death of Schwann cells (SCs) leading to loss of myelin sheath and nerve destruction [3, 4]. TT shows only few lesions, and bacilli can rarely be identified. Patients with poor T-cell mediated immunity show the lepromatous-type leprosy (LL). LL leads to massive bacterial load inside host cells specially SCs and macrophages [3, 5–7]. The lesions of TT and LL types are named as T-lep and L-lep lesions, respectively, but damage of the nerves is observed in most of the cases of both types [7].

Lepromatous leprosy exhibits multiple lesions of the skin, eyes, nerves, and lymph nodes, which are characterized by tumor-like accumulations of foamy macrophages. The foamy macrophages are fully packed with lipid droplets (LDs) and contain high numbers of lepra bacilli. These aggregations of foamy macrophages expand slowly and disfigure the body of the host [8]. Foamy macrophages were described first by the pathologist Rudolph Virchow [9] and despite these lipid-laden cells are a hallmark of lepromatous leprosy, only recently a few researchers started to elucidate the molecular biology of these cells. In the last ten years, it has become evident that M. tuberculosis induces the formation of foamy macrophages, a process which appears to be a key event in both sustaining persistent bacteria and release of infectious bacilli [10]. Moreover host lipids from LDs are regarded as substantial nutrient source for mycobacteria during infection. Middlebrook already demonstrated in the late 1940s that mycobacterial growth in vitro was enhanced by supplementation with oleic acid [11]. Wheeler et al. reported insufficient fatty acid synthase activity to support growth in M. leprae. This finding supported the hypothesis that M. leprae scavenges lipids from the host cell [12]. Over the last years, it has become evident that survival and persistence of M. tuberculosis is critically dependent on lipid body formation. Furthermore, lipid body formation seems to be the prerequisite for transition of M. tuberculosis to the dormant state. This goes along with the important observation that sputum from tuberculosis patients contains lipid body-laden bacilli [13, 14].

For a great number of mycobacterial species, intracellular lipid bodies were reported [15–25]. In the dormant state, lipids from lipid bodies appear to be the primary carbon source for M. tuberculosis in vivo. For M. tuberculosis, several bacterial genes are upregulated during the dormant state and have been reported to be involved in lipid metabolism such as diacylglycerol acyltransferase (tgs1), lipase (lipY), and isocitrate lyase (icl) [26, 27].

M. leprae has a small genome (3.2 Mb). The obligate intracellular organism shows a moderate genome degradation and several genes are absent when compared with other mycobacterial species. Due to the gene loss, M. leprae is strongly dependent on the host for basic metabolic functions [3, 28]. Macrophages infected with M. leprae contain oxidized host lipids and it has been observed that M. leprae upregulates 13 host lipid metabolism genes in T-lep lesions and 26 in L-lep lesions. The oxidized lipids inhibit innate immune responses and thus seem to be an important virulence factor for the organism [29].

This paper highlights the importance of the LDs as one of the most unique determinant for M. lepraepersistence and virulence. The formation of LDs in M. leprae infected cells has been compared to LDs formation in macrophages infected with M. tuberculosis. Thereafter, we will focus on the interplay between the LDs of the host and the pathogen.

M. LEPRAE INDUCES LIPID DROPLETS FORMATION IN THE HOST CELL

M. leprae infects preferentially macrophages and SCs [6]. A typical feature of lepromatous leprosy is the survival and replication of M. leprae stored within the LDs in the enlarged phagosome of histiocytes. Due to strong immune response against M. leprae and arrested bacilli growth in T-lep lesions, the occurrence of foamy cells of M. leprae-infected macrophages/SCs is a marker only in L-lep lesions [5, 9]. In L-lep lesions, the expression of lipoprotein and fatty acid metabolism, including lipases genes, was 4-fold greater than in T-lep lesions which supports the previous observation [29]. Lipid droplets are thought to be an important nutrient source for the bacillus. A major concern in leprosy is peripheral neuropathy. The damage to nerves of the peripheral nervous system is caused by the infection of SCs by M. leprae. In LL nerve biopsies, highly infected SCs also contain LDs and show a foamy appearance, such as Virchow cells found in dermal lesions [30].

The biology of these foamy cells has been characterized poorly until now. Neither the origin nor nature of the lipids has been elucidated yet. Only recently, in vitro studies by Mattos could show that M. leprae induces the formation of LDs in human SCs [5]. Moreover, the group found that LDs are promptly recruited to bacterial phagosomes. In SCs, LDs recruiting by bacterial phagosomes depends on cytoskeletal reorganization and PI3 K signaling but is independent of TLR2 bacterial sensing [5].

Important markers for the lipid accumulation in adipocytes or macrophages are lipid-droplet-associated proteins such as adipose differentiation-related protein ADRP and perilipin, which play essential roles in lipid-droplet formation [31]. After phagocytosis of live M. leprae, ADRP expression is constantly upregulated in human monocytes. ADRP and perilipin are localized at the phagosomal membrane (Figure 1) [31].

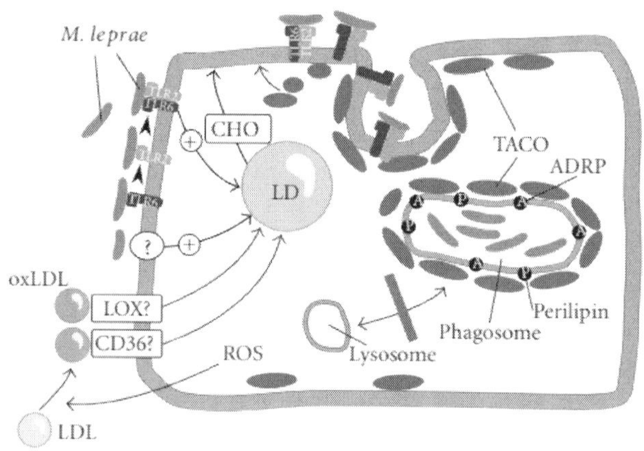

Figure 1: Basic mechanisms of lipid droplet biogenesis in macrophages infected with M. leprae. TLR mediated uptake of M. leprae: M. leprae attaches to

TLR2 and TLR6. Heterodimerization of TLR2 and TLR6 induces downstream signalling and subsequent cholesterol accumulation by LDs formation [32, 33]. In SCs, TLR6, but not TLR2, is essential for M. leprae-induced LDs biogenesis [5]. Cholesterol accumulates at the site of mycobacterial entry and promotes mycobacterial uptake. Cholesterol also recruits TACO from the plasma membrane to the phagosome [34]. TACO prevents phagosome-lysosome fusion and promotes intracellular survival [35, 36]. Uptake by scavenger receptors (proven for M. tuberculosis, hypothetical for M. leprae): reactive oxygen species might oxidize low-density lipoprotein (LDL) to oxLDL, which is thought to be subsequently bound and taken up by scavenger receptors CD36 and LOX1. A: ADRP; CHO: cholesterol; P: perilipin. Unknown mechanisms for LDs induction are indicated with a question mark.

Receptor-Mediated LDs Formation

Mycobacterium bovis Bacillus Calmette-Guérin (BCG) and M. leprae are recognised by the Toll-like receptors (TLR) TLR6 and TLR2 [32, 37] and induces the formation of foamy macrophages [38]. For BCG it has been shown that surface molecule lipoarabinomannan (LAM) binds to TLR2. [39].

M. leprae association to macrophages is mediated by binding of the bacteria to TLR2 and TLR6. Heterodimerization of TLR2 and TLR6 leads to downstream signalling and subsequent LDs formation (Figure 1) [32, 33]. Macrophage association is not dependent on binding to TLR2 or TLR6. Neither a TLR2$^{-/-}$ nor TLR6$^{-/-}$ knockout macrophage shows reduced binding to M. leprae. This suggests that both TLR2 and TLR6 can bind M. leprae alone, or/and the presence of other receptors, binding to M. leprae. The TLR2$^{-/-}$ or TLR6$^{-/-}$knockout macrophages do not also completely abolish LD formation, but show only reduced LDs formation [32]. This suggests the presence of additional signalling pathways for LDs formation. In contrast TLR6, but not TLR2, is essential for M. leprae-induced LDs biogenesis in SCs [5].

At least one additional factor is required for general phagocytosis, in mycobacterial infection. Members of the CD36 family have to be shown to be required for uptake of mycobacteria [40]. Macrophages infected with M. tuberculosis show an increased surface expression of the type 1 scavenger receptors CD36 and LOX1, which facilitate the uptake of oxidized host macromolecules including OxLDL (Figure 1) [41]. These findings are consistent with the observation that genes for ADRP and CD36 are upregulated in L-lep lesions, accumulated with LD enriched macrophages

[29, 32, 42]. Macrophages generate and release reactive oxygen species (ROS) during activation of the respiratory burst upon infection with pathogenic bacteria. Oxidative stress results not only in damage to cellular structures but also to oxidation of fatty acids, such as low density lipoproteins (OxLDL) in granulomas. OxLDL binding to LOX-1 also increases reactive oxygen species (ROS) formation [41, 43]. The binding of OxLDL to type 1 scavenger receptors CD36 and LOX1 induces increased surface expression of both receptors, leading to uptake of OxLDL [41, 43, 44]. In addition, CD36 increases the uptake of M. tuberculosis by macrophages [40]. The increased rate of OxLDL uptake results in the accumulation of oxidized lipids, which finally leads to the formation of foamy macrophages (Figure 1) [41]. M. tuberculosis and M. leprae benefit from the accumulated OxLDL in the infected macrophage. OxLDL-laden lung macrophages show enhanced replication of intracellular M. tuberculosis compared to macrophages loaded with nonoxidized LDL [41]. The presence of oxidized phospholipids in M. leprae infected macrophages downregulates the innate immune response and contributes to pathogenesis [29]. Moreover, scavenger receptor-deficient phagocytes are characterized by a reduced intracellular bacterial survival and a lower cytokine response [45].

Lipid Composition in the Foamy Macrophage

In 1863, Virchow described lipid-laden macrophages. The lipids in these foamy cells form droplets and surround also M. leprae within the phagososomes [46, 47]. This lipid layer or capsule forms a characteristic electron-transparent zone. In contrast to M. tuberculosis, lipid inclusions (lipid bodies) seem to be rather exceptional in M. leprae [47]. The electron-transparent zone contains mycoserosoic acids of phthiocerol dimycocerosates as well as phenolic glycolipids [48, 49]. Brennan reported the full characterization of three phenol-phthiocerol triglycosides by M. leprae [46]. It has been postulated that many of these same molecules together with phosphatidylinositol mannosides and phospholipids are released from the cell wall after synthesis, forming the capsule-like region [6]. The dominant lipid in the cell wall which gives M. lepraeimmunological specificity is phenolic glycolipid-1 (PGL-1). Phenolic glycolipid 1 has been isolated in relatively high concentrations from purified bacteria and from M. leprae infected tissues [50]. PGL-1 is thought to be a major component of the capsule in M. leprae and constitutes an important interface between bacteria and host [51]. It has been suggested that PGL-1 is involved in the interaction of M. leprae with the laminin of SCs, thus PGL-1 might play a role in peripheral nerve-bacillus interactions [52]. Moreover, phenolic glycolipids seem to be involved in the stimulation of suppressor T-cells in lepromatous leprosy [53]. Daniel et al. demonstrated that M. tuberculosis inside foamy macrophages imports fatty acids derived from host TAG and incorporates them intact into bacterial TAG, which is accumulated in lipid bodies [54].

Recently, it was reported that also LDs from M. leprae infected SCs and macrophages accumulate mainly host-derived lipids, such as oxidized phospholipids [29]. BODIPY stains infected SCs, indicating that LDs contain neutral lipids, such as triacylglycerols (TAG) [5].

Regulation of Host-Signaling by M. leprae Induced-Lipid Droplets

In M. leprae infected-cells, LDs show high ability to fuse further to form giant LDs. In most mammalian cells, dynein and extracellular signal-regulated kinase-2 (ERK2) facilitate LDs move along the cytoskeletal network [55]. Extracellular signal-regulated kinase-1/2 is one of the most upregulated signals after M. leprae induced-SCs demyelination and nerve injury [56]. Using NIH 3T3 cells, Andersson et al. demonstrated that ERK2 is an essential signal in LDs formation [57] and also seem to be crucial for the phospholipase D1- (PLD1-) mediated increases of LDs. The ERK2 mechanism is mediated by phosphorylation of dynein, which finally upregulates its association in ADRP-containing LDs [57].

Lipid droplets induction in all type of cells associates with lipid droplets protein adipophilin (ADFP) [58]. The classic model of LDs postulates that LDs formation take place in the outer leaflet of the endoplasmic reticulum (ER) membrane (the place of TAG synthesis). In this formation, the function of AFDP has been proposed in fatty acid absorption and transport. Cruz et al. showed that ADFP is overexpressed in L-lep lesions [29]. Furthermore, the authors also pinpointed similarity in overall observations to one of the important metabolic disease atherosclerosis. Similar observation in atherosclerosis showed that ADFP is the most expressed LD-associated protein found in macrophages and that the accumulation of LDs in lesional macrophages is associated with ADFP aggregation [58].

Several reports have supported the inflammatory role of macrophages and SCs, where LDs induction by these cells associated with modulation in production of inflammatory mediators [59–63]. Moreover, increased LDs formation accompanied by PGE2 and transforming growth factor-β

(TGF-β1) synthesis in BCG infected macrophage while strong PGE2 was seen in foamy macrophages during M. tuberculosis infection [64, 65]. Cytokines, chemokines, and growth factors seem to be correlated with LDs formation by M. leprae. For instance, Persson et al. found that IL-1β and TNF-α inflate foamy macrophages formation [66]. M. lepraeinfection significantly increased NF-κB, TNF-α, and IL-1β expression in NOD1- and NOD2-transfected macrophages cells [67]. Another example, IL-10 and IL-12 secretion correlated with bacilli-LD induction in L-lep lesions [32, 68]. Schwann cells are able to produce several type of cytokines, including IL-1, IL-6, IL-8, IL-10, IL-12, PGE2, TGF-β, and TNF-α [37, 69, 70]. PGE2 and IL-10 overexpression is associated with clog of IL-12 and NO production in M. leprae-infected SCs [32, 68]. In late infection state, M. leprae impede the expression of TNF signaling and induce expression of 9-O-acetyl GD3 as smart strategy to avoid the cellular apoptosis and facilitate its persistence inside the favored cell niche [37, 67, 71, 72]. Obviously, lipid biosynthesis and cytokine production as the whole immune response to M. leprae infection are well connected in SCs [5]. Innate immune response in infected SCs seems to be dependent on LDs accumulation and TLR2/TLR6 signaling. Using a nonsteroidal anti-inflammatory drug and fatty acids synthesis inhibitors, both LDs formation and innate immune responses completely abrogated, supporting the role of bacilli-LDs induction and immune responses [32, 37]. Another modulation observed is downregulation of Th1 responses and bactericidal activity through overproduced PGE2 where TLR2 and TLR6 heterodimerization plays important role [32]. In general, cytokine profile in both lesions well linked with TLRs function. M. leprae modified-TLRs function stimulates monocytes and macrophages in TT and LL disease type, but in L-lep lesions the adaptive response was debilitated more than what was seen in T-lep lesions [6]. Despite the strong evidences that LDs induction is associated with modulation of cytokines, it is not known whether cytokines signaling functions within LDs themselves.

LIPID METABOLISM IN M. LEPRAE AND M. TUBERCULOSIS

Due to the genome reduction M. leprae is strongly dependent on catalytic activities provided by the host [3, 28]. Macrophages infected with M. leprae contain oxidized host lipids and it has been observed that M. lepraeupregulates 13 host lipid metabolism genes in T-lep lesions and 26 in L-lep lesions [29]. But still it is not clear what carbon source M. leprae uses under aerobic or anaerobic conditions. The genome contains genes for utilization of sugars, such as glycolysis, tricarboxylic acid cycle (TCA), and monophosphate shunt (HMP) pathways. The M. leprae genome contains all genes for a functional glyoxylate cycle, which would enable the organism to use acetyl-CoA from the β-oxidation of lipids. M. tuberculosis derives much of its energy from the degradation of host-derived lipids [73, 74]. Neutral lipids are hydrolyzed by lipases or esterases, yielding fatty acids for energy generation and anabolism of membrane phospholipids. The genome of M. tuberculosis H37Rv contains twenty genes designated as putative lipases (lipA to W, except K and S) [74, 75]. In the M. lepraegenome, only 2 lipase genes (lipG, lipU) were found, suggesting a gene loss in comparison to M. tuberculosis. But recently it has been found that in M. tuberculosis genome only six Lip expressed-enzymes showed reasonable hydrolase activity for long-chain triacylglycerols (LipY, LipC, LipL, LipX, LipK, and LipG). LipY (Rv3097c) has the highest activity, compared to all Lip enzymes. LipG and LipU from M. leprae are homologous with LipG and LipU from M. tuberculosis and show sequence identities of 72% and 79%, respectively. The lipases LipG and LipU from M. tuberculosis show very low and no activity with long-chain triacylglycerols as substrates [76]. M. tuberculosis LipY is suspected to be a major functional lipase, which utilizes stored triacylglycerols (TAG) during dormancy and reactivation of the pathogen [76, 77]. LipY shows only a weak similarity with M. leprae LipU (23% identity). In summary, it appears that M. leprae uses different lipases for the hydrolysis of fatty acids than M. tuberculosis. All M. tuberculosis and their M. leprae homologs enzymes involved in lipid metabolism in lipid bodies are summarized in Table 1.

Table 1: Comparison between the enzymes involved in TAG metabolism in M. tuberculosisand M. bovis BCG and their homologous in M. leprae. ND, not determined

Gene	Enzyme	Diacylglycerol acyltransferase activity (in vitro)	Influence on lipid bodies/TAG accumulation	M. leprae homologue	M. leprae homologue-diacylglycerol acyltransferase activity (in vitro)	Reference
Rv3130c(tgs1)	DGAT	+	Δtgs1decreases TAG accumulation	ML1244	ND	[26, 78]
Rv3734c(tgs2)	DGAT	+	ND	ML1244	ND	[26, 78]
Rv3234c (tgs3)	DGAT	+	ND	ML1244	ND	[26, 78]
Rv3088(tgs4)	DGAT	+	ND	ML1244	ND	[26, 78]
Rv1760	DGAT	+	ND	ML1244	ND	[26, 78]
Rv2285	DGAT	+	ND	ML1244	ND	[26, 78]
Rv3804c (85A)	DGAT	+	Overexpression increases production of lipid bodies	ML0097 (85A)	ND	[79]
BCG1489c[Rv1428c]	DGAT	ND	ΔBCG1489c reduces TAG accumulation	ML2427c	ND	[80]
BCG3153c(tgs1) [Rv3130c]	DGAT	ND	ΔBCG3153c reduces TAG accumulation	ML1244	ND	[77]
BCG3794c(tgs2) [Rv3734c]	DGAT	ND	ΔBCG3794c reduces TAG accumulation	ML1244	ND	[80]

Gene	Enzyme	TAG-hydrolyzing activity (in vitro)	Influence on lipid bodies/TAG accumulation	M. lepraehomologue	M. leprae homologue-TAG-hydrolyzing activity (in vitro)	Reference
Rv3097c(lipY)	Lipase/esterase	+	ΔlipY reduces TAG hydrolysis. Overexpression increases TAG hydrolysis	ML0314c (lipU) ML1053 ML1183c	ND	[76]
Rv1399c(lipH)	Lipase/esterase	+	ND	ML0314c (lipU)	ND	[75]
BCG1721[Rv1683]	Lipase/esterase	+	+	ML1346	ND	[80]
Rv0183	Lipase/esterase	Hydrolyzes only monoacylglycerides	ND	ML2603	ND	[81]
Rv2224c	Lipase/esterase	ND	ND	ML1633c	ND	[82]
Gene	Enzyme	Isocitrate cleavage (in vitro)	Influence on lipid bodies/TAG accumulation	M. lepraehomologue	M. leprae homologue-isocitrate cleavage (in vitro)	Reference
Rv0467 (icl) Rv1916(icl2)	isocitrate lyase	+	ND (but the Δicl, Δicl2 strain shows no intracellular replication)	ML1985c (aceA)	ND	[83–87]

M. tuberculosis can grow on fatty acids as sole carbon source and it has been demonstrated that fatty acid oxidation is important for survival of the pathogen in the lungs of mice [83, 84]. Fatty acids are oxidized via the β-oxidation cycle and the glyoxylate shunt, to replenish TCA cycle intermediated during growth [88]. The β-oxidation cycle consists of five biochemical reactions, where one molecule acetyl-CoA of the fatty acid is split off per cycle. The genome of M. tuberculosis encodes around 100 genes, designated as fad genes (fatty acid degradation) with putative roles in the β-oxidation of fatty acids. While E. coli has only one enzyme for each step of the β-oxidation cycle, M. tuberculosis seems to have several backup enzymes for each reaction [89]. M. leprae has approximately one-third as many potential fad enzymes. However, there are three times more FadD acyl-CoA synthases than there are FadE acyl-CoA dehydrogenases, whereas these are predicted in equal amounts in M. tuberculosis [74]. The initial step of β-oxidation is the formation of acyl-CoA from free fatty acids and coenzyme A and is catalyzed by acyl-CoA synthase. The Acyl-CoA synthase Rv1683 is suspected to be essential for TAG hydrolysis and growth [80, 90].

Even though M. leprae genome contains less necessary β-oxidation cycle genes than M. tuberculosis, transcripts analysis revealed active expression of acyl-CoA metabolic enzymes including echA1 (ML0120, putative enoyl-CoA hydratase), echA12 (ML1241, possible enoyl-CoA hydratase), fadA2 (ML2564, acetyl-CoA-acetyltransferase), fadB2 (ML2461, 3-hydroxyacyl-CoA dehydrogenase), fadD19 (ML0352, acyl-CoA synthase), fadD26 (ML2358, fatty acid-CoA-ligase), fadD29 (ML0132, probable fatty-acid-CoA synthetase), fadD28(ML0138, possible fatty-acid-CoA synthase), fadE25 (ML0737, probable acyl-CoA dehydrogenase), and fadE5(ML2563, acyl-CoA dehydrogenase) [91, 92]. Combing observations from leprosy lesions, this gives strong evidence that host lipids provide the main carbon and energy sources for M. leprae during infection.

Together with malate synthase, isocitrate lyase (ICL) is the key enzyme of the glyoxylate cycle that catalyzes the cleavage of isocitrate to glyoxylate and succinate [88, 93]. The M. leprae genome contains a gene, coding for an isocitrate lyase, aceA. AceA is upregulated in both M. leprae-infected nude mouse and human lesions. [91]. The amino acid sequence of aceA (ML1985c) shows 80% identity with its homologue from M. tuberculosis ICL2 (Rv1915/1916). A second icl gene is not present in M. leprae, as observed in M. tuberculosis. This finding is of particular interest, because two M. tuberculosis isocitrate lyases, icl and icl2, are jointly required for in vivo growth and virulence [83, 84]. Deletion of icl1 or icl2 has little effect on bacterial growth in macrophages [83]. So far the evidence indicates that M. leprae aceA might play a slightly different role in both isocitrate lyases in M. tuberculosis.

M. LEPRAE HAS A REDUCED NUMBER OF TRIACYLGLYCEROL SYNTHASE GENES

For persistent M. tuberculosis, TAGs seem to be the major carbon and energy source. Biosynthesis of TAG consists of the sequential esterification of the glycerol moiety with fatty acyl-residues by various acyltransferases. The fatty acids are supplied by de novo biosynthesis or β-oxidation. Esterification occurs via sequential acylation of the sn-1,2 and 3 positions of glycerol-3-phosphate, and removal of the phosphate group before the last acylation step. The terminal reaction is the esterification of diacylglycerol (DAG) with acyl-CoA by an diacylglycerol acyltransferase (Figure 2) [94]. Bacteria do not contain DGATs but only bifunctional wax ester synthase/acyl-CoA: diacylglycerol acyltransferases (WS/DGAT). WS/DGATs mediate next to TAG formation the synthesis of waxes by esterification of acyl-CoA with alcohol [95]. The genome of M. tuberculosiscodes for 15 genes which contain the highly conserved putative active site motif of WS/DGATs (HHxxxDG). These genes were designated as "tgs," triacylglycerol synthases, but have only a weak sequence similarity to other WS/DGAT sequences. All 15 expressed mycobacterial tgs proteins show diacylglycerol acyltransferase activity and tgs1 has the highest activity of all enzymes [26]. Tgs1 appears to be a major contributor to TAG synthesis in M. tuberculosis so far [26, 78]. And moreover, two homologue proteins to tgs1 and tgs2 (BCG3153c and BCG3794c) and another poorly characterized acyltransferase (BCG1489c) were found to be exclusively associated to lipid bodies. Disruption of the tgs genes BCG3153c, BCG3794c, and BCG1489c reduces TAG accumulation during hypoxia-induced nonreplicating state, revealing that the enzymes are involved in TAG synthesis during latency and pathogenicity [80].

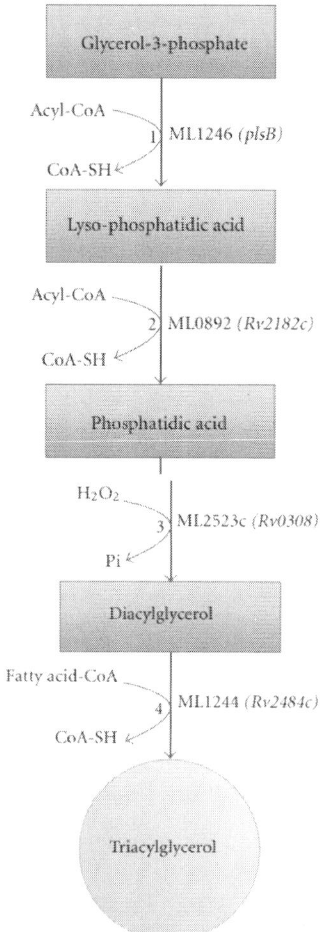

Figure 2: Schematic representation of putative genes involved in TAG biosynthesis in M. leprae. As a precursor, glycerol-3-phosphate further processed by sequential acylation steps, until last the transesterification of diacylglycerol (DAG) to triacylglycerol (TAG). Reaction 1 mediated by glycerol-3-phosphate acyltransferase, 1-acylglycerol-3-phosphate acyltransferase mediates reaction 2, phosphatidc acid phosphatase processes reaction 3, and the last reaction 4 is mediated by acyl-CoA: diacylglycerol acyltransferase enzyme.

Ten of the 15 tgs genes in M. tuberculosis are located adjacent or proximal to 11 lip genes that are annotated as probable phospholipases or lipases-esterases carboxylesterases. Some tgs genes may be cotranscribed with neighboring lip genes and may synthesize triacylglycerols from the released fatty acids from the host [26]. Lip gene products may be important for utilization of TAGs during dormancy and upon reactivation after dormancy.

The tgs gene Rv0221 is located near LipC (Rv0220), lipW (Rv0217c), acyl-CoA synthetase (Rv0214), acyl-CoA dehydrogenase (Rv0215c), and an integral membrane acyltransferase (Rv0228). This clustering of genes of the fatty acid metabolism suggests that these genes may be cotranscribed and may release fatty acid from host TAG, carry out the transport of fatty acids and finally catalyze the resynthesis of TAGs in the pathogen. Rv0221 and LipC have to be shown to be catalytical active [26, 76].

The M. leprae genome shows only one predicted gene product which has a significant degree of identity to any tgs enzymes from M. tuberculosis [26]. The tgs gene product ML1244 shows 72% identity to Rv2484c from M. tuberculosis. Rv2484c is located next to a carboxylesterase lipQ, (Rv2485c), a probable glycerol-3-phosphate acyltransferase, (Rv2482c), a lysophosphatidic acid acyltransferase-like protein (Rv2483c), and a probable enoyl-CoA hydratase (Rv2486). The gene cluster of lipid metabolism genes suggests a possible involvement of the gene products in the synthesis of TAG [26]. A few tgs genes (Rv3234c, Rv3233c, Rv2285, and Rv1425) are located proximally to lipoproteins, which may serve as donors or acceptors of fatty acids [26].

Ag85A, a mycoly transferase that is known to catalyze the formation of the cord factor was recently found to have additional DGAT activity [79]. The homologue gene of M. tuberculosis Ag85A is expressed in M. leprae. Although, mycoly transferase 85 complex genes (A, B, and C) transcripts were one of the major secreted group during mycobacterium infection of the mouse, Ag85A is one of the most genes shown to be relatively highly expressed either in infected nude mouse or human skin lesions [91].

CHOLESTEROL ACCUMULATION IN THE HOST CELL AND UTILIZATION BY M. LEPRAE

Macrophages, infected with M. leprae, alter the lipid metabolism. Especially an increased accumulation of cholesterol and cholesterol esters has been found [42]. The increased cholesterol level was also associated with a lower level of esterase activity in infected macrophages [42].

Mattos et al. recently showed by high-performance thin-layer chromatography that the amount of cholesterol increases in SCs upon infection with M. leprae [5]. Cholesterol utilization was also identified to be required for mycobacterial persistence [96]. In 2008, Pandey and Sassetti found that M. tuberculosis can grow using cholesterol as a primary carbon source and that the mce4 transporter is required for cholesterol uptake.

While the M. tuberculosis genome contains four homologous mce operons, mce1-mce4, which are thought to encode lipid transporters [96, 97], M. leprae genome encodes only one operon (mce1). All M. leprae five mcegenes were overexpressed during intracellular growth in mouse and human biopsies [91, 92]. This observation suggests that the intracellular bacilli population induces cholesterol uptake of the infected cell and subsequently uses the stored cholesterol as carbon and energy source.

Cholesterol is also essential for survival of mycobacteria in macrophages. Cholesterol accumulates at the site of mycobacterial entry in macrophages and promotes mycobacterial uptake. For M. tuberculosis and M. leprae it has been shown that cholesterol mediates the recruitment of TACO from the plasma membrane to the phagosome [34]. TACO, also termed as coronin-1A (CORO1A), is a coat protein that prevents phagosome-lysosome fusion and thus degradation of mycobacteria in lysosomes [34, 35]. The entering of mycobacteria at cholesterol-rich domains of the plasma membrane and their subsequent uptake in TACO-coated phagosomes promotes intracellular survival [35, 36].

CONCLUSIONS

Studying M. leprae metabolisms during infection is as complex as the choice of invasion mechanisms traffic and as their multifarious factors for entry are. In this scenario, most of publications emphasize that leprosy maintains metabolic remodeling between both host and pathogen. A hallmark of intracellular infection is the formation of foamy host cells. Several results support the view that M. leprae actively catabolizes fatty acids for energy and produces a wide array of secretory proteins. Huge gaps remain in our knowledge of the structure and function of M. leprae lipid bodies, as the host metabolism required alterations to switch the organism to the utilization of lipids as alternate carbon sources. Investigating the metabolic aspects of M. leprae especially LDs formation is an important topic to learn more about metabolic pathways that could potentially be exploited for developing novel drugs against leprosy and tuberculosis.

ACKNOWLEDGMENTS

This work was supported by NOPERSIST (EC Grant Agreement no. 232188) grant from European Commission, Seventh Framework Programme (FP7) to M. Singh.

REFERENCES

1. D. S. Ridley and W. H. Jopling, "Classification of leprosy according to immunity. A five-group system," International Journal of Leprosy and Other Mycobacterial Diseases, vol. 34, no. 3, pp. 255–273, 1966.
2. WHO, "WHO Leprosy Today," WHO, 2010.
3. C. L. Cosma, D. R. Sherman, and L. Ramakrishnan, "The secret lives of the pathogenic mycobacteria," Annual Review of Microbiology, vol. 57, pp. 641–676, 2003.
4. D. M. Scollard, "Endothelial cells and the pathogenesis of lepromatous neuritis: insights from the armadillo model," Microbes and Infection, vol. 2, no. 15, pp. 1835–1843, 2000.
5. K. A. Mattos, F. A. Lara, V. G. C. Oliveira et al., "Modulation of lipid droplets by Mycobacterium leprae in Schwann cells: a putative mechanism for host lipid acquisition and bacterial survival in phagosomes," Cellular Microbiology, vol. 13, no. 2, pp. 259–273, 2011.
6. D. M. Scollard, L. B. Adams, T. P. Gillis, J. L. Krahenbuhl, R. W. Truman, and D. L. Williams, "The continuing challenges of leprosy," Clinical Microbiology Reviews, vol. 19, no. 2, pp. 338–381, 2006.
7. V. D. Vissa and P. J. Brennan, "Impact of the Mycobacterium leprae genome sequence on leprosy research," in Genomics of GC-Rich Gram-Positive Bacteria, A. Danchin, Ed., Band 2, pp. 85–118, Caister Academic Press, 2002.
8. D. S. S. Raphael Rubin and R. Emanuel, Eds., Rubin's Pathology: Clinicopathologic Foundations of Medicine, Lippincott Williams & Wilkins, 2007.
9. R. Virchow, Die Krankhaften Geschwülste, August Hirschwald, Berlin, Germany, 1863.
10. D. G. Russell, P. J. Cardona, M. J. Kim, S. Allain, and F. Altare, "Foamy macrophages and the progression of the human tuberculosis granuloma," Nature Immunology, vol. 10, no. 9, pp. 943–948, 2009.
11. R. J. Dubos and G. Middlebrook, "Media for tubercle bacilli," The American Review of Tuberculosis, vol. 56, no. 4, pp. 334–345, 1947.
12. P. R. Wheeler, K. Bulmer, and C. Ratledge, "Enzymes for biosynthesis de novo and elongation of fatty acids in mycobacteria grown in host cells: is Mycobacterium leprae competent in fatty acid biosynthesis?" Journal of General Microbiology, vol. 136, no. 1, pp. 211–217, 1990.

13. O. Neyrolles, R. Hernández-Pando, F. Pietri-Rouxel et al., "Is adipose tissue a place for Mycobacterium tuberculosis persistence?" PLoS ONE, vol. 1, no. 1, article e43, 2006.
14. N. J. Garton, S. J. Waddell, A. L. Sherratt et al., "Cytological and transcript analyses reveal fat and lazy persister-like bacilli in tuberculous sputum," PLoS Medicine, vol. 5, no. 4, article e75, 2008.
15. K. L. Burdon, "Fatty material in bacteria and fungi revealed by staining dried, fixed slide preparations," Journal of Bacteriology, vol. 52, no. 6, pp. 665–678, 1946.
16. G. Knaysi, J. Hillier, and C. Fabricant, "The cytology of an avian strain of Mycobacterium tuberculosisstudied with the electron and light microscopes," Journal of Bacteriology, vol. 60, no. 4, pp. 423–447, 1950.
17. W. B. Schaefer and C. W. Lewis Jr., "Effect of oleic acid on growth and cell structure of mycobacteria," Journal of Bacteriology, vol. 90, no. 5, pp. 1438–1447, 1965.
18. E. M. Brieger and A. M. Glauert, "Electron microscopy of the leprosy bacillus: a study of submicroscopical structure," Tubercle, vol. 37, no. 3, pp. 195–206, 1956.
19. G. R. Gale and H. H. Mclain, "Effect of ethambutol on cytology of Mycobacterium smegmatis," Journal of Bacteriology, vol. 86, pp. 749–756, 1963.
20. R. L. Whitehouse, P. C. Wong, and F. L. Jackson, "Ultrastructure of Mycobacterium lepraemurium," International Journal of Leprosy and Other Mycobacterial Diseases, vol. 39, no. 2, pp. 151–163, 1971.
21. R. E. Weir, P. E. M. Fine, S. Floyd et al., "Comparison of IFN-γ responses to mycobacterial antigens as markers of response to BCG vaccination," Tuberculosis, vol. 88, no. 1, pp. 31–38, 2008.
22. L. Barksdale and K. S. Kim, "Mycobacterium," Bacteriological Reviews, vol. 41, no. 1, pp. 217–372, 1977.·
23. P. Draper, "The anatomy of the mycobacteria," in The Biology of the Mycobacteria, C. Ratledge and J. Stanford, Eds., pp. 9–52, Academic Press, London, UK, 1982.
24. R. M. van Boxtel, R. S. Lambrecht, and M. T. Collins, "Effect of polyoxyethylene sorbate compounds (Tweens) on colonial morphology, growth, and ultrastructure of Mycobacterium paratuberculosis," Acta Pathologica, Microbiologica et Immunologica Scandinavica, vol. 98,

no. 10, pp. 901–908, 1990.

25. N. J. Garton, H. Christensen, D. E. Minnikin, R. A. Adegbola, and M. R. Barer, "Intracellular lipophilic inclusions of mycobacteria in vitro and in sputum," Microbiology, vol. 148, no. 10, pp. 2951–2958, 2002.·
26. J. Daniel, C. Deb, V. S. Dubey et al., "Induction of a novel class of diacylglycerol acyltransferases and triacylglycerol accumulation in Mycobacterium tuberculosis as it goes into a dormancy-like state in culture," Journal of Bacteriology, vol. 186, no. 15, pp. 5017–5030, 2004.
27. P. Peyron, J. Vaubourgeix, Y. Poquet et al., "Foamy macrophages from tuberculous patients' granulomas constitute a nutrient-rich reservoir for M. tuberculosis persistence," PLoS Pathogens, vol. 4, no. 11, Article ID e1000204, 2008.
28. D. A. Hagge, S. Oby Robinson, D. Scollard, G. McCormick, and D. L. Williams, "A new model for studying the effects of Mycobacterium leprae on Schwann cell and neuron interactions," The Journal of Infectious Diseases, vol. 186, no. 9, pp. 1283–1296, 2002.
29. D. Cruz, A. D. Watson, C. S. Miller et al., "Host-derived oxidized phospholipids and HDL regulate innate immunity in human leprosy," The Journal of Clinical Investigation, vol. 118, no. 8, pp. 2917–2928, 2008.·
30. C. K. Job, "Mycobacterium leprae in nerve lesions in lepromatous leprosy. An electron microscopic study," Archives of Pathology, vol. 89, no. 3, pp. 195–207, 1970.
31. K. Tanigawa, K. Suzuki, K. Nakamura et al., "Expression of adipose differentiation-related protein (ADRP) and perilipin in macrophages infected with Mycobacterium leprae," FEMS Microbiology Letters, vol. 289, no. 1, pp. 72–79, 2008.
32. K. A. Mattos, H. D'Avila, L. S. Rodrigues et al., "Lipid droplet formation in leprosy: Toll-like receptor-regulated organelles involved in eicosanoid formation and Mycobacterium leprae pathogenesis," Journal of Leukocyte Biology, vol. 87, no. 3, pp. 371–384, 2010.
33. D. M. Underhill, "Toll-like receptors: networking for success," European Journal of Immunology, vol. 33, no. 7, pp. 1767–1775, 2003.
34. K. Suzuki, F. Takeshita, N. Nakata, N. Ishii, and M. Makino, "Localization of CORO1A in the macrophages containing Mycobacterium leprae," Acta Histochemica et Cytochemica, vol. 39, no. 4, pp. 107–112, 2006.

35. J. Gatfield and J. Pieters, "Essential role for cholesterol in entry of mycobacteria into macrophages," Science, vol. 288, no. 5471, pp. 1647–1650, 2000.
36. P. K. Anand and D. Kaul, "Downregulation of TACO gene transcription restricts mycobacterial entry/survival within human macrophages," FEMS Microbiology Letters, vol. 250, no. 1, pp. 137–144, 2005.
37. K. A. Mattos, V. G. C. Oliveira, H. D'Avila et al., "TLR6-driven lipid droplets in Mycobacterium leprae-infected Schwann cells: immunoinflammatory platforms associated with bacterial persistence," The Journal of Immunology, vol. 187, no. 5, pp. 2548–2558, 2011.
38. H. D'Avila, R. C. N. Melo, G. G. Parreira, E. Werneck-Barroso, H. C. Castro-Faria-Neto, and P. T. Bozza, "Mycobacterium bovis bacillus Calmette-Guérin induces TLR2-mediated formation of lipid bodies: intracellular domains for eicosanoid synthesis in vivo," The Journal of Immunology, vol. 176, no. 5, pp. 3087–3097, 2006.
39. P. E. Almeida, A. R. Silva, C. M. Maya-Monteiro et al., "Mycobacterium bovis bacillus Calmette-Guérin infection induces TLR2-dependent peroxisome proliferator-activated receptor γ expression and activation: functions in inflammation, lipid metabolism, and pathogenesis," The Journal of Immunology, vol. 183, no. 2, pp. 1337–1345, 2009.
40. J. A. Philips, E. J. Rubin, and N. Perrimon, "Drosophila RNAi screen reveals CD36 family member required for mycobacterial infection," Science, vol. 309, no. 5738, pp. 1251–1253, 2005.
41. G. S. Palanisamy, N. M. Kirk, D. F. Ackart et al., "Uptake and accumulation of oxidized low-density lipoprotein during Mycobacterium tuberculosis infection in guinea pigs," PLoS One, vol. 7, no. 3, Article ID e34148, 2012.
42. I. G. Kurup and P. R. Mahadevan, "Cholesterol metobolism of macrophages in relation to the presence of Mycobacterium leprae," Journal of Biosciences, vol. 4, no. 3, pp. 307–316, 1982.
43. R. Yoshimoto, Y. Fujita, A. Kakino, S. Iwamoto, T. Takaya, and T. Sawamura, "The discovery of LOX-1, its ligands and clinical significance," Cardiovascular Drugs and Therapy, vol. 25, no. 5, pp. 379–391, 2011.·
44. S. Collot-Teixeira, J. Martin, C. McDermott-Roe, R. Poston, and J. L. McGregor, "CD36 and macrophages in atherosclerosis," Cardiovascular Research, vol. 75, no. 3, pp. 468–477, 2007.

45. A. Leelahavanichkul, A. V. Bocharov, R. Kurlander et al., "Class B scavenger receptor types I and II and CD36 targeting improves sepsis survival and acute outcomes in mice," The Journal of Immunology, vol. 188, no. 6, pp. 2749–2758, 2012.
46. P. J. Brennan, "Mycobacterium leprae—the outer lipoidal surface," Journal of Biosciences, vol. 6, no. 5, pp. 685–689, 1984.
47. C. A. Fisher and L. Barksdale, "Cytochemical reactions of human leprosy bacilli and mycobacteria: ultrastructural implications," Journal of Bacteriology, vol. 113, no. 3, pp. 1389–1399, 1973.
48. I. Sakurai and O. K. Skinsnes, "Lipids in leprosy. 2. Histochemistry of lipids in human leprosy," International Journal of Leprosy and Other Mycobacterial Diseases, vol. 38, no. 4, pp. 389–403, 1970.
49. G. Kaplan, W. C. van Voorhis, E. N. Sarno et al., "The cutaneous infiltrates of leprosy. A transmission electron microscopy study," The Journal of Experimental Medicine, vol. 158, no. 4, pp. 1145–1159, 1983.·
50. S. W. Hunter and P. J. Brennan, "A novel phenolic glycolipid from Mycobacterium leprae possibly involved in immunogenicity and pathogenicity," Journal of Bacteriology, vol. 147, no. 3, pp. 728–735, 1981.
51. L. Guenin-Macé, R. Siméone, and C. Demangel, "Lipids of pathogenic Mycobacteria: contributions to virulence and host immune suppression," Transboundary and Emerging Diseases, vol. 56, no. 6-7, pp. 255–268, 2009.
52. V. Ng, G. Zanazzi, R. Timpl et al., "Role of the cell wall phenolic glycolipid-1 in the peripheral nerve predilection of Mycobacterium leprae," Cell, vol. 103, no. 3, pp. 511–524, 2000.
53. G. Puzo, "The carbohydrate- and lipid-containing cell wall of mycobacteria, phenolic glycolipids: structure and immunological properties," Critical Reviews in Microbiology, vol. 17, no. 4, pp. 305–327, 1990.
54. J. Daniel, H. Maamar, C. Deb, T. D. Sirakova, and P. E. Kolattukudy, "Mycobacterium tuberculosis uses host triacylglycerol to accumulate lipid droplets and acquires a dormancy-like phenotype in lipid-loaded macrophages," PLoS Pathogens, vol. 7, no. 6, Article ID e1002093, 2011.
55. D. J. Murphy, "The dynamic roles of intracellular lipid droplets: from

archaea to mammals," Protoplasma, vol. 249, no. 3, pp. 541–585, 2012.

56. A. Rambukkana, "Usage of signaling in neurodegeneration and regeneration of peripheral nerves by leprosy bacteria," Progress in Neurobiology, vol. 91, no. 2, pp. 102–107, 2010.
57. L. Andersson, P. Boström, J. Ericson et al., "PLD1 and ERK2 regulate cytosolic lipid droplet formation," Journal of Cell Science, vol. 119, no. 11, pp. 2246–2257, 2006.
58. B. H. J. Chang and L. Chan, "Regulation of Triglyceride Metabolism. III. Emerging role of lipid droplet protein ADFP in health and disease," American Journal of Physiology, vol. 292, no. 6, pp. G1465–G1468, 2007.
59. E. Spierings, T. de Boer, B. Wieles, L. B. Adams, E. Marani, and T. H. M. Ottenhoff, "Mycobacterium leprae-specific, HLA class II-restricted killing of human Schwann cells by $CD4^+$ Th1 cells: a novel immunopathogenic mechanism of nerve damage in leprosy," The Journal of Immunology, vol. 166, no. 10, pp. 5883–5888, 2001.
60. R. B. Oliveira, M. T. Ochoa, P. A. Sieling et al., "Expression of Toll-like receptor 2 on human Schwann cells: a mechanism of nerve damage in leprosy," Infection and Immunity, vol. 71, no. 3, pp. 1427–1433, 2003.
61. M. T. Talaue, V. Venketaraman, M. H. Hazbón et al., "Arginine homeostasis in J774.1 macrophages in the context of Mycobacterium bovis BCG infection," Journal of Bacteriology, vol. 188, no. 13, pp. 4830–4840, 2006.
62. R. C. Melo, H. D'Avila, H. C. Wan, P. T. Bozza, A. M. Dvorak, and P. F. Weller, "Lipid bodies in inflammatory cells: structure, function, and current imaging techniques," The Journal of Histochemistry and Cytochemistry, vol. 59, no. 5, pp. 540–556, 2011.
63. Y. Yuan, P. Li, and J. Ye, "Lipid homeostasis and the formation of macrophage-derived foam cells in atherosclerosis," Protein & Cell, vol. 3, no. 3, pp. 173–181, 2012.
64. H. D'Avila, N. R. Roque, R. M. Cardoso, H. C. Castro-Faria-Neto, R. C. N. Melo, and P. T. Bozza, "Neutrophils recruited to the site of Mycobacterium bovis BCG infection undergo apoptosis and modulate lipid body biogenesis and prostaglandin E2 production by macrophages," Cellular Microbiology, vol. 10, no. 12, pp. 2589–2604, 2008.
65. J. Rangel Moreno, I. Estrada García, M. de la Luz García Hernández,

D. Aguilar Leon, R. Marquez, and R. Hernandez Pando, "The role of prostaglandin E2 in the immunopathogenesis of experimental pulmonary tuberculosis," Immunology, vol. 106, no. 2, pp. 257–266, 2002.

66. J. Persson, J. Nilsson, and M. W. Lindholm, "Interleukin-1beta and tumour necrosis factor-alpha impede neutral lipid turnover in macrophage-derived foam cells," BMC Immunology, vol. 9, article 70, 2008.·

67. A. M. Skoff, R. P. Lisak, B. Bealmear, and J. A. Benjamins, "TNF-alpha and TGF-beta act synergistically to kill Schwann cells," Journal of Neuroscience Research, vol. 53, no. 6, pp. 747–756, 1998.

68. M. Yamamura, K. Uyemura, R. J. Deans et al., "Defining protective responses to pathogens: cytokine profiles in leprosy lesions," Science, vol. 254, no. 5029, pp. 277–279, 1991.

69. R. P. Lisak, D. Skundric, B. Bealmear, and S. Ragheb, "The role of cytokines in Schwann cell damage, protection, and repair," The Journal of Infectious Diseases, vol. 176, supplement 2, pp. S173–S179, 1997.·

70. J. L. Rutkowski, G. F. Tuite, P. M. Lincoln, P. J. Boyer, G. I. Tennekoon, and S. L. Kunkel, "Signals for proinflammatory cytokine secretion by human Schwann cells," Journal of Neuroimmunology, vol. 101, no. 1, pp. 47–60, 1999.

71. Y. Qin, C. Cheng, H. Wang, X. Shao, Y. Gao, and A. Shen, "TNF-α as an autocrine mediator and its role in the activation of Schwann cells," Neurochemical Research, vol. 33, no. 6, pp. 1077–1084, 2008.

72. V. T. Ribeiro-Resende, M. L. Ribeiro-Guimarães, R. M. Lemes et al., "Involvement of 9-O-Acetyl GD3 ganglioside in Mycobacterium leprae infection of Schwann cells," The Journal of Biological Chemistry, vol. 285, no. 44, pp. 34086–34096, 2010.

73. P. R. Wheeler and C. Ratledge, "Metabolism of Mycobacterium tuberculosis," in Tuberculosis: Pathogenesis, Protection, and Control, B. R. Bloom, Ed., pp. 353–385, Washington, DC, USA, 1994.

74. S. T. Cole, K. Eiglmeier, J. Parkhill et al., "Massive gene decay in the leprosy bacillus," Nature, vol. 409, no. 6823, pp. 1007–1011, 2001.

75. S. Canaan, D. Maurin, H. Chahinian et al., "Expression and characterization of the protein Rv1399c from Mycobacterium tuberculosis: a novel carboxyl esterase structurally related to the HSL family," European Journal of Biochemistry, vol. 271, no. 19, pp. 3953–

3961, 2004.

76. C. Deb, J. Daniel, T. D. Sirakova, B. Abomoelak, V. S. Dubey, and P. E. Kolattukudy, "A novel lipase belonging to the hormone-sensitive lipase family induced under starvation to utilize stored triacylglycerol in Mycobacterium tuberculosis," The Journal of Biological Chemistry, vol. 281, no. 7, pp. 3866–3875, 2006.
77. K. L. Low, P. S. Rao, G. Shui et al., "Triacylglycerol utilization is required for regrowth of in vitro hypoxic nonreplicating Mycobacterium bovis bacillus Calmette-Guerin," Journal of Bacteriology, vol. 191, no. 16, pp. 5037–5043, 2009.
78. T. D. Sirakova, V. S. Dubey, C. Deb et al., "Identification of a diacylglycerol acyltransferase gene involved in accumulation of triacylglycerol in Mycobacterium tuberculosis under stress," Microbiology, vol. 152, no. 9, pp. 2717–2725, 2006.
79. A. A. Elamin, M. Stehr, R. Spallek, M. Rohde, and M. Singh, "The Mycobacterium tuberculosis Ag85A is a novel diacylglycerol acyltransferase involved in lipid body formation," Molecular Microbiology, vol. 81, no. 6, pp. 1577–1592, 2011.
80. K. L. Low, G. Shui, K. Natter et al., "Lipid droplet-associated proteins are involved in the biosynthesis and hydrolysis of triacylglycerol in Mycobacterium bovis bacillus Calmette-Guérin," The Journal of Biological Chemistry, vol. 285, no. 28, pp. 21662–21670, 2010.
81. K. Côtes, R. Dhouib, I. Douchet et al., "Characterization of an exported monoglyceride lipase from Mycobacterium tuberculosis possibly involved in the metabolism of host cell membrane lipids," Biochemical Journal, vol. 408, no. 3, pp. 417–427, 2007.
82. S. Lun and W. R. Bishai, "Characterization of a novel cell wall-anchored protein with carboxylesterase activity required for virulence in Mycobacterium tuberculosis," The Journal of Biological Chemistry, vol. 282, no. 25, pp. 18348–18356, 2007.
83. E. J. Muñoz-Elías and J. D. McKinney, "Mycobacterium tuberculosis isocitrate lyases 1 and 2 are jointly required for in vivo growth and virulence," Nature Medicine, vol. 11, no. 6, pp. 638–644, 2005.
84. J. D. McKinney, K. Höner zu Bentrup, E. J. Muñoz-Elías et al., "Persistence of Mycobacterium tuberculosisin macrophages and mice requires the glyoxylate shunt enzyme isocitrate lyase," Nature, vol. 406, no. 6797, pp. 735–738, 2000.

85. V. Sharma, S. Sharma, K. Hoener zu Bentrup et al., "Structure of isocitrate lyase, a persistence factor of Mycobacterium tuberculosis," Nature Structural & Molecular Biology, vol. 7, no. 8, pp. 663–668, 2000.·
86. R. Kumar and V. Bhakuni, "Mycobacterium tuberculosis isocitrate lyase (MtbIcl): role of divalent cations in modulation of functional and structural properties," Proteins, vol. 72, no. 3, pp. 892–900, 2008.
87. K. Höner zu Bentrup, A. Miczak, D. L. Swenson, and D. G. Russell, "Characterization of activity and expression of isocitrate lyase in Mycobacterium avium and Mycobacterium tuberculosis," Journal of Bacteriology, vol. 181, no. 23, pp. 7161–7167, 1999.
88. S. A. Ensign, "Revisiting the glyoxylate cycle: alternate pathways for microbial acetate assimilation," Molecular Microbiology, vol. 61, no. 2, pp. 274–276, 2006.
89. K. J. Williams, H. I. Boshoff, N. Krishnan, J. Gonzales, D. Schnappinger, and B. D. Robertson, "The Mycobacterium tuberculosisβ-oxidation genes echA5 and fadB3 are dispensable for growth in vitro and in vivo," Tuberculosis, vol. 91, no. 6, pp. 549–555, 2011.
90. M. Stehr, A. A. Elamin, and M. Singh, "Cytosolic lipid droplets formed during infection by bacterial and viral pathogens," Microbes and Infection, vol. 14, no. 13, pp. 1227–1237, 2012.
91. D. L. Williams, M. Torrero, P. R. Wheeler et al., "Biological implications of Mycobacterium leprae gene expression during infection," Journal of Molecular Microbiology and Biotechnology, vol. 8, no. 1, pp. 58–72, 2004.
92. T. Akama, K. Tanigawa, A. Kawashima, H. Wu, N. Ishii, and K. Suzuki, "Analysis of Mycobacterium leprae gene expression using DNA microarray," Microbial Pathogenesis, vol. 49, no. 4, pp. 181–185, 2010.·
93. H. Kornberg, "Krebs and his trinity of cycles," Nature Reviews Molecular Cell Biology, vol. 1, no. 3, pp. 225–228, 2000.
94. H. M. Alvarez and A. Steinbüchel, "Triacylglycerols in prokaryotic microorganisms," Applied Microbiology and Biotechnology, vol. 60, no. 4, pp. 367–376, 2002.
95. R. Kalscheuer and A. Steinbüchel, "A novel bifunctional wax ester synthase/acyl-CoA: diacylglycerol acyltransferase mediates wax ester and triacylglycerol biosynthesis in Acinetobacter calcoaceticus ADP1,"

The Journal of Biological Chemistry, vol. 278, no. 10, pp. 8075–8082, 2003.

96. A. K. Pandey and C. M. Sassetti, "Mycobacterial persistence requires the utilization of host cholesterol," Proceedings of the National Academy of Sciences of the United States of America, vol. 105, no. 11, pp. 4376–4380, 2008.

97. L. I. Klepp, M. A. Forrellad, A. V. Osella et al., "Impact of the deletion of the six mce operons in Mycobacterium smegmatis," Microbes and Infection, vol. 14, no. 7-8, pp. 590–599, 2012.

CHAPTER 8

HERBAL MEDICINE FOR THE TREATMENT OF OBESITY: AN OVERVIEW OF SCIENTIFIC EVIDENCE FROM 2007 TO 2017

Yanfei Liu[1,2], Mingyue Sun[2], Hezhi Yao[2], Yue Liu[3], and Rui Gao[2]

[1]Graduate School of Beijing University of Chinese Medicine, Beijing 100029, China
[2]Institute of Clinical Pharmacology, Xiyuan Hospital of China Academy of Chinese Medical Sciences, Beijing 100091, China
[3]Cardiovascular Diseases Center, Xiyuan Hospital of China Academy of Chinese Medical Sciences, Beijing 100091, China

ABSTRACT

Obesity is a very common global health problem, and it is known to be linked to cardiovascular and cerebrovascular diseases. Western medical treatments for obesity have many drawbacks, including effects on monoamine neurotransmitters and the potential for drug abuse and dependency. The safety of these medications requires improvement. Herbal medicine has been used for treatment of disease for more than 2000 years, and it has prov-

Citation: Yanfei Liu, Mingyue Sun, Hezhi Yao, Yue Liu, and Rui Gao. Herbal Medicine for the Treatment of Obesity: An Overview of Scientific Evidence from 2007 to 2017, Evidence-Based Complementary and Alternative Medicine. Volume 2017, Article ID 8943059, 17 pages. https://doi.org/10.1155/2017/8943059

en efficacy. Many studies have confirmed that herbal medicine is effective in the treatment of obesity, but the mechanisms are not clear. This article will discuss the possible effects and mechanisms of herbal medicine treatments for obesity that have been reported in the past decade.

INTRODUCTION

Obesity is a metabolic disorder characterized by an excess accumulation of fat in the body due to energy intake exceeding energy expenditure [1]. Obesity is an increasingly common phenomenon all over the world. Body mass index (BMI) is the most commonly used measure to evaluate the degree of obesity. In 2016, the AACE (the American Association of Clinical Endocrinologists) released new diagnostic criteria of obesity based on BMI combined with obesity-related complications (see Table 1) [2]. The latest study, which analyzed data from 68.5 million persons between 1980 and 2015, found that a total of 107.7 million children and 603.7 million adults were obese in 2015 [3]. Obesity has become a worldwide epidemic, and the trend is becoming increasingly serious. Obesity is an independent risk factor for metabolic syndrome; major medical problems associated with the development of hypertension, type 2 diabetes (T2DM), dyslipidemia, sleep apnea, and respiratory disorders; and ultimately life-threatening cardiovascular disease (CVD), stroke, and certain types of cancer [4–6].

Table 1: The new definition of obesity from the AACE* [2]

Diagnosis	Body mass index (BMI)	Clinical component (complications)
Overweight	≥25–29.9	No complications
Obesity stage 0	≥30	No complications
Obesity stage 1	≥25	One or more mild-to-moderate complications
Obesity stage 2	≥25	One or more severe complications

*AACE: American Association of Clinical Endocrinologists

The number of obese patients is increasing globally [7]. Reducing body weight by lifestyle alteration is advisable, but sometimes drug intervention is necessary [8]. Obesity drugs can be divided into five categories: central appetite suppressants, digestion and absorption blockers, metabolic promoters, obesity gene product inhibitors, and other drugs for the treatment of obesity [9]. However, the weight loss drugs prescribed in conventional medicine induce many adverse reactions, primarily effecting monoamine neurotransmitters, and causing drug abuse or dependence [10]. For example, sibutramine has been reported to commonly cause adverse events, including dry mouth, insomnia, anorexia, constipation, formation of thrombi, and neurological symptoms [11, 12]. Surgery is commonly used in morbidly obese patients (BMI $\geq 40\,kg/m^2$) or in patients with comorbidities, such as hypertension, diabetes, and obstructive sleep apnea [13]. Common surgical complications include infection, postoperative anastomotic fistula, deep vein thrombosis, and long-term complications such as anemia and malnutrition [14, 15]. Given the dangers of obesity and the shortcomings of western medicine, alternative treatments should be further investigated. This article examines the potential role of herbal medicines in the treatment of obesity and summarizes the scientific evidence reported from 2007 to 2017.

METHODS

The PubMed and Web of Science were searched for studies published from 2007 to 2017 on humans or animals. The search terms were "obesity," "obese," or "antiobesity" and "herbal medicine," "plant," "plant medicine," or "Chinese medicine" without narrowing or limiting search items. Relevant publications with available abstracts and titles were reviewed by two reviewers.

The Clinical Trials (https://clinicaltrials.gov/) and Chinese Clinical Trial Registry (http://www.chictr.org.cn/) databases were searched for registered clinical trials of herbal medicine and obesity. The search terms were "obesity" or "obese" and "herbal medicine," "plant," "plant medicine," or "Chinese medicine."

RESULTS

The Role of Herbal Medicine in Treating Obesity: Evidence from Human Studies

Eighteen randomized controlled trials (RCTs) (sample size > 50 cases) [16–33] published from 2007 to 2017 were included. Studies of herbal medicine interventions for obesity that had no obvious effects were excluded. The contents of the included 18 published RCTs are shown in Table 2. Analysis of these studies found that the maximum number of subjects was only 182, and the sample size is small. The age of the subjects ranged from 18 to 79 years.

The studies were performed in many different populations. Eleven studies [16–18, 20, 21, 23, 25, 28–30, 33] mentioned complications, including hypertension, impaired glucose tolerance, spleen hypofunction, excessive sweating, nonalcoholic fatty liver disease, hyperlipidemia, and metabolic syndrome. Of the 18 studies, 6 were completed by Chinese researchers, and the remaining 12 were from Japan, Australia, Canada, USA, Russia, France, Indonesia, Korea, Indian, Thailand, and Italy. Thus, herbal medicine interventions for obesity are being studied in more countries than China. The outcome of each study varied and could be roughly divided into the following categories: (1) change in body weight: a significant decrease in body weight occurred following treatment with xin-ju-xiao-gao-fang (XJXGF, compound of rhubarb, Coptis, semen cassiae, andCitrus aurantium), yellow pea fiber, bofu-tsusho-san (compound of Radix Platycodi, Gypsum Fibrosum, talcum, Paeoniae, Scutellariae, and Glycyrrhizae), RCM-104 (compound of Camellia sinensis, flos sophorae, and semen cassiae), pistachio, Satiereal®, Monoselect Camellia (containing green tea extract: GreenSelect® Phytosome®), or Nigella sativa; (2) BMI: a significant decrease in body fat occurred following treatment with xin-ju-xiao-gao-fang, bofu-tsusho-san, RCM-104, Linggui Zhugan Decoction (compound of poria, Macrocephalae, Radix Glycyrrhizae, Ramulus Cinnamomi, and Radix Atractylodis), Pu'er tea, pistachio, or Monoselect Camellia; (3) waist or hip circumference: there was a significant decrease in waist or hip circumferences treat with the following herbal medicine from six studies: xin-ju-xiao-gao-fang, Pu'er tea, Satiereal, Catechin enriched green tea, West African Plant (Irvingia gabonensis), and Cissus quadrangularis(Irvingia gabonensis); (4) food intake: two studies, of RCM-104 and yellow pea fiber, referred to the influence of traditional Chinese medicine on food intake, but data were not provided; (5) other effects: homeostatic model assessment-insulin resistance (HOMA-IR), homeostatic model assessment-β cell function (HOMA-β), glycated hemoglobin (HbA1c), blood pressure (BP), quality of life, fasting insulin (FINS), and fasting plasma glucose (FPG) were detected in these trials; (6) evaluating these eighteen clinical studies based on Jadad score: it was found that the overall quality of these clinical studies is low. Of three studies, the Jadad score was 4, and the remaining studies scored below 4 scores. We found 16 registered clinical trials (see Table 3) from https://clinicaltrials.gov/ and http://www.chictr.org.cn/, and the recruiting locations vary from China and Korea to United States and Portugal, which will provide greater scientific insight into the treatment of obesity by herbal medicine all over the world.

Table 2: Published randomized controlled trials of herbal medicines for treatment of obesity in humans from 2007 to 2017

Number	Authors/year	Targets	Age (years)	Name of herb or formula	Jadad score	Dose/duration	Groups	Main outcomes	Weight (kg) before treatment	Weight (kg) after treatment	Adverse events
(1)	Lambert et al. (2016) [16]	n=53	18–70	Yellow pea fiber	4	15 g/day, 12 weeks	I: yellow pea fiber C: placebo	Body weight↓ HbAlc↓ Food intake↓ Plasma glucose↓ Satiety↑, regulating gut microbiota	92.3 ± 4.1	91.5 ± 4.0	No reports
(2)	Azushima et al, (2015) [17]	n=106	20–79	Bofu-tsusho-san (Platycodi, Gypsum Fibrosum, talcum, Paeoniae, Scutellariae, Glycyrrhizae)	2	7.5 g/day, 24 weeks	I: compound C: placebo	Body weight↓ BMI↓ HbA1c↓ BP↓	82.5 ± 16.4	78.3 ± 17.9	Gastric irritation, constipation, elevation of serum hepatic enzyme level
(3)	Zhou et al. (2014) [18]	n=140	18–60	Xin-ju-xiao-gao-fang (rhubarb, Coptis, semen cassia, Citrus aurantium)	3	170 mL/day, 24 weeks	I: full-dose C: low-dose	Body weight↓ BMI↓ Waist Circumference↓ Fasting insulin↓ HOMA-IR↓ HOMA-β↓	91.8 ± 13.4	Reduce 3.6 ± 0.5	Skin rash

(4)	Lenon et al. (2012) [19]	n=117	18–60	RCM-104 (Camellia sinensis, flos sophorae, semen cassiae)	4	500 mg granule extract/day, 12 weeks	I: compound C: placebo	Body weight↓ BMI↓ Body fat↓ Food intake↓	99.5 ± 15.1	98 ± 15.4	Nausea, headache
(5)	Ke et al. (2012) [20]	n=95	25–70	Linggui Zhugan Decoction (poria Macrocephalae, Radix Glycyrrhizae, Ramulus Cinnamomi, Radix Atractylodis)	2	Dose is unknown twice a day, 24 weeks	I: Linggui Zhugan Decoction combined with short-term very low calorie diets C: basic weight-reduction treatment	BMI↓, SBP↓ DBP↓, FPG↓ 2hPG↓, TC↓ TG↓	99.5 ± 15.1	/	Fatigue, hunger, dizziness
(6)	Chu et al. (2011) [21]	n=90	18–70	Pu'er tea	3	4 cap/day, 12 weeks	I: extract C: placebo	BMI↓ Waist-hip ratio↓ TC↓ TG↓ FBG↓ PG2h↓	/	/	Diarrhea
(7)	Li et al. (2010) [22]	n=59	20–65	Pistachio	3	53 g/day 12 weeks	I: pistachio C: pretzels	Body weight↓ BMI↓	86.1 ± 1.4	82.4 ± 1.6	No reports
(8)	Abidov et al. (2010) [23]	n=151	/	Xanthigen (brown marine algae fucoxanthin, pomegranate seed oil)	4	2.4 mg/day, 16 weeks	I: extract C: control	Body weight↓ Body liver fat content↓	92.5 ± 1.5	88.2 ± 1.9	No adverse effects
(9)	Gout et al. (2010) [24]	n=60	25–45	Satiereal, (Crocus sativusL. extract)	1	176.5 mg/day, 8 weeks	I: extract C: placebo	Body weight↓ BMI↓ Waist circumference↓	73.2 ± 1.1	72.2 ± 1.2	Nausea, diarrhea, reflux

(10)	Datau et al. (2010) [25]	n=50	30–45	Nigella sativa	2	750 mg twice daily, 12 weeks	I: extract C: flour	Body weight↓ SBP↓	77.1 ± 4.9	72.6 ± 5.4	No reports
(11)	Di Pierro et al. (2009) [26]	n=100	25–60	Green tea extract	1	50 mg/day 90 days	I: hypocaloric diet + extract C: hypocaloric diet	Body weight↓ BMI↓	96.1 ± 18.0	82.3 ± 15.3	No reports
(12)	Wang et al. (2009) [27]	n=182	18–55	Catechin enriched green tea	2	458 mg, 468 mg, 886 mg/day, 90 days	I: extract C: Placebo	Body weight↓ Waist circumference↓ Total body fat↓	71.1 ± 11.9	69.9 ± 12.1	No adverse events
(13)	He et al. (2009) [28]	n=102	18–65	Oolong tea	1	8 g/6 weeks	I: extract C: control	Body weight↓ Subcutaneous fat content↓, TC↓, TG↓	Men: 79.7 ± 6.7 Women: 70.2 ± 6.8	Women: 67.8 ± 6.7 Men: 70.2 ± 6.8	No adverse events
(14)	Ngondi et al. (2009) [29]	n=102	19–50	West African Plant (Irvingia gabonensis)	2	150 mg/10 weeks	I: extract C: placebo	Body weight↓, body fat↓ Waist circumference↓ TC↓, LDL-C↓ Leptin levels↑	97.9 ± 9.1	85.1 ± 3.1	Headache, sleep difficulty, intestinal flatulence
(15)	Oben et al. (2008) [30]	n=72	21–44	Cissus quadrangularis, Irvingia gabonensis	2	Unknown/twice daily/10 weeks	I: Cissus quadrangularis or Cissus quadrangularis-Irvingia gabonensiscombination; C: placebo	Body weight↓, body fat↓, waist size↓, FBG↓, LDL-C↓ TC↓	99.8 ± 13.5	88.0 ± 3.2	Headache, lack of sleep, gas

(16)	Roong-pisuthipong et al. (2007) [31]	n=50	18–75	Garcinia atro-viridis	2	1.15 grams of Garcinia atroviridis/day/8 weeks	I: diet + extract C: diet	Body weight↓ BMI, ↓ triceps skin fold thick-ness↓	69 ± 1	Reduce 2.8 ± 0.1	No adverse events
(17)	Kuriyan et al. (2007) [32]	n=50	28–53	Caralluma fimbriata	3	1 g/60 days	I: weight loss program + ex-tract; C: weight loss program	Hunger levels Body weight↓ BMI↓, body fat↓ energy intake↓	79.5 ± 16.9	77.2 ± 8.6	Abdomi-nal dis-tention, flatu-lence, consti-pation, gastritis
(18)	Oben et al. (2007) [33]	n=168	19–54	Cissus quadran-gularis	2	300, 1028 mg/8 weeks	I: two-extract formulation: CQR-300, CORE; C: placebo	Body weight↓ Body fat↓ glu-cose↓ HDL-C↑	118.6 ± 3.8	113.8 ± 2.5	No reports

Table 3: The registered clinical trials of herbal medicines for the treatment of obesity from 2007 to 2017

Num-ber	Trial number status	Conditions and dosage	Objectives	Interventions	Outcomes	Num-ber of subjects (age/sex*)	Date*	Recruit-ing study locations
(1)	ChiCTR-IOR-15007587(pending)	Obesity (/)	To evaluate the effective-ness and safety of the empirical formula—Xiere Huazhuo Formula of Chi-nese Medicine Professor—Ding Xueping in obesity treatment	I: Xiere Huazhuo granule; C: orlistat	1: weight, body fat distribution, blood lipid, insulin resistance HOMA-2, adipo-kines	48 (18–65/F-M)	Decem-ber 14, 2015	China

(2)	NCT00383058(completed)	Obesity (/)	To examine whether extract of the green tea is effective on obese women	I: the extract of green tea; C: placebo	Body mass index, body weight, glucose, cholesterol, LDL, HDL, triglyceride	100 (16–60/F)	September 29, 2009	China
(3)	NCT02605655(completed)	Metabolic syndrome X (1 g/day for 3 months)	To determine whether the Chinese formula AMP-1915 has effect on metabolic syndrome (MS) in MS patients	I: AMP-1915 (Astragalus, Radix Puerariae, Cortex Mori); C: placebo	FBG, plasma lipid levels, plasma insulin concentration, body weight, HbA1c	60 (40–65/F-M)	April 1, 2015	China
(4)	NCT01142076(completed)	Overweight (170 mL/day, 24 weeks)	To examine the treatment of adiposity (stagnation of QI causing phlegm retention)	I: Xinju Xiaogao Prescription; C: placebo	Waistline, BMI	140 (16–80/F-M)	March 1, 2011	China
(5)	NCT02651454(recruiting)	Obesity (6 g, three times a day, 12 weeks)	To investigate the efficacy and safety of Daesiho-tang and Taeeumjowi-tang on Korean obese women with metabolic syndrome risk factors	I: Daesiho-tang: Jowiseungcheung-tang; C: placebo	Body weight, body fat percentage, fat mass, waist circumference, body mass index, lipid profile	120 (18–65/F)	January 5, 2016	Korea
(6)	NCT02337933(completed)	Metabolic syndrome X (150 mg, once a day, 12 weeks)	To evaluate the effect of ursolic acid on the insulin sensitivity and metabolic syndrome	I: ursolic acid; C: placebo	Total insulin sensitivity, waist circumference, fasting glucose levels, body weight, BMI	24 (30–60/F-M)	September 1, 2015	Mexico
(7)	NCT01724099(recruiting)	Obesity (3 times per day, 12 weeks)	To evaluate the effect of Euiiyin-tang on obese patients	I: Euiiyin-tang; C: placebo	Weight, C-reactive protein, blood pressure, blood glucose, waist/hip ratio	160 (18–65/F)	November 2, 2012	Korea

(8)	NCT02929849(ongoing)	Obesity (300 mg, 500 mg/day)	To determine whether an herb with known alpha-glucosidase inhibitor properties (Salacia chinensis, SC), affecting postprandial appetite ratings and glucose indices in overweight/obese individuals	I: Salacia chinensis; C: placebo	Appetite ratings, glucose indices, gut hormones	59 (20–59/F-M)	August 16, 2016	United States
(9)	NCT01778257(completed)	Obesity (mate extract (3150 mg/day), 12 weeks)	To evaluate efficacy and safety of mate extracts on decrement of body and abdominal fat in obese subjects	I: mate extract; C: placebo	Body and abdominal fat, weight, BMI, waist and hip circumference	30 (19–65/F-M)	March 1, 2012	Korea
(10)	NCT01709955(completed)	Obesity (750 mg of Glucomannan in capsule form)	To determine if the herb, Glucomannan, is an effective nonpharmacological appetite suppressant for overweight or class I obese patients	I: Glucomannan, C: placebo	Weight	43 (21–60/F-M)	July 1, 2011	United States
(11)	NCT00502658(completed)	Overweight, obesity (dose is unknown, 12 weeks)	To evaluate the effect of dietary supplements (shakes and supplements) and personal energy tracking device to promote and maintain healthy weight	I: dietary supplement containing vitamins, minerals, and herbs; C: dietary supplement	Body weight, biophotonic scanner	120 (18–65/F-M)	December 1, 2007	United States
(12)	NCT00823381(completed)	Obesity, metabolic syndrome (75 mg once a day)	To evaluate the effects of the antioxidant "resveratrol" to a diet intervention (calorie restriction)	I: resveratrol, C: placebo	Insulin sensitivity, body composition, blood lipid levels	58 (35–70/F)	December 1, 2013	United States

(13)	NCT02613715(completed)	Overweight and obesity (250 mL of blackberry juice)	To evaluate the bioavailability of blackberry juice anthocyanins in normal weight and overweight/ obese adults	I: blackberry juice C: blackberry juice with 12% ethanol	Plasma concentrations of anthocyanins and anthocyanin metabolites	18 (18–40/ M-F)	June 2015	Portugal
(14)	NCT01705093(unknown)	Childhood obesity; cardiovascular disease (50 g of flavonoid-rich freeze-dried strawberry powder)	To verify if strawberry intake can lead to improvements in select measures of cardiovascular function in overweight and obese adolescent males	I: flavonoid-rich freeze-dried strawberry powder C: macronutrient-matched control powder	Vascular functionmeasured by peripheralarterial tonometry	25 (14–18/M)	August 2012	United States
(15)	NCT01138930(unknown)	Polycystic ovary syndrome; obesity (1.5 g daily for 3 months)	To examine the effect of berberine metabolic and hormonal parameters and insulin resistance in obese patients with polycystic ovary syndrome	I: berberine; C: placebo	Body insulin action, Weight, waist/hip circumference, OGTT	120 (18–35/F)	June 7, 2010	China
(16)	NCT01471275(unknown)	Type 2 diabetes mellitus; obesity; high triglycerides (15 g each time, twice a day, with boiled water)	Evaluate the safety and efficacy of Jiang Tang Tiao Zhi decoction in treatment of obesity with type 2 diabetes, dyslipidemia	I: Jiang Tang Tiao Zhi decoction; C: metformin	Glycosylated hemoglobin, waistline, triglycerides, liver function	450 (30–65/ F-M)	November 14, 2011	China

*If the status is completed, the date is completion date; others are registration date; F = female; M = men.

The Role of Herbal Medicine in Treating Obesity: Evidence from Animal Studies

In this section, we will summarize the known effects and mechanisms of action of single herbs and their components or extracts in animal models of obesity (see Table 4 and Figure 1).

Table 4: The antiobesity effects of single herbs and their components or extracts in animal models

Herb	Animal	Model	Dose/administration/time	Effects	Components	Reference
Rhizoma coptidis	Mice	High-fat diet-fed C57BL/6J mice	Berberine (200 mg/kg) oral gavage/6 weeks	Visceral adipose↓ Weight↓ Blood glucose↓ Lipid levels↓	Berberine	[34, 35]
Panax ginseng C. A. Mey	Mice	High-fat diet-fed mice	20 mg/kg/intragastric administration/3 weeks	Body weight↓ Food intake↓ Blood glucose↓ TC↓ TG↓	Ginsenoside Rg1 Ginsenoside Rb1 Ginsenoside Rg3	[36–38]
Radix Lithospermi	Rat	High-fat diet-fed db/db mice	Acetylshikonin extract (100, 300, or 900 mg/kg)/intragastric administration/6 weeks; db/db mice: acetylshikonin (540 mg/kg/day) oral/8 weeks	Body weight↓ FFA↓ TG↓ inhibited Differentiation Fat accumulation, Food intake↓	Acetylshikonin; shikonin	[39–42]
Ephedra sinica Stapf.	Male ICR mice	High-fat diet-fed	Diet containing 5% Ephedra/oral gavage/6 weeks	Body weight ↓ Fasting glucose↓ HDL-C↑	Ephedra	[43, 44]
Rheum palmatum L.	Mice	Obese mice	Mice: chrysophanic acid (5 mg/kg/day)/oral gavage/16 weeks	Body weight↓ TG↓ HDL-↑, TC↓ Food intake↓	Chrysophanic acid	[45–47]
Green tea	Mice	Diet-induced obese male C57BL/6J mice	0.25% (w/w) GT extract/oral gavage/12 weeks	Body weight↓ Energy intake HOMA-IR↓, TG ↓, TC↓ FFA↓	Catechin	[27, 48]
Astragalus membranaceus (Fisch.) Bunge	Mice	db/db diabetic mice	Radix Astragali (2 g/kg/day)/oral gavage/12 weeks	Body weight ↓ Food intake ↓ HDL-C↑	Astragalosides I Astragalosides II	[49, 50]
Carthamus tinctorius L	Rat	High-fat diet-induced obese rats	Saffron extract and crocin at concentrations of 40 and 80 mg/kg/day oral/8 weeks	Food intake, relative liver weight	Saffron, crocin	[51, 52]

Ganoderma lucidum (Leyss. ex Fr.) Karst	Mice	ob/ob mice	100 μL water extract of G. lucidum mycelium/ intragastric gavage/8 weeks	Inflammation endotoxemia ↓ Insulin resistance ↓ Regulated lipogenic gene expression.	Saffron	[53]
Tripter-ygium wil-fordii Hook. f	Mice	High-fat diet-induced obese db/ db or ob/ob mice	Celastrol (100 μg/ kg)/ intraperitoneally injection/3 weeks	HFD: food intake ↓ Energy expendi-ture ↑ Body weight↓ db/db or ob/ ob mice: body weights, lean mass, and fat percentage were not affected	Celastrol	[54]

10 Evidence-Based Complementary and Alternative Medicine

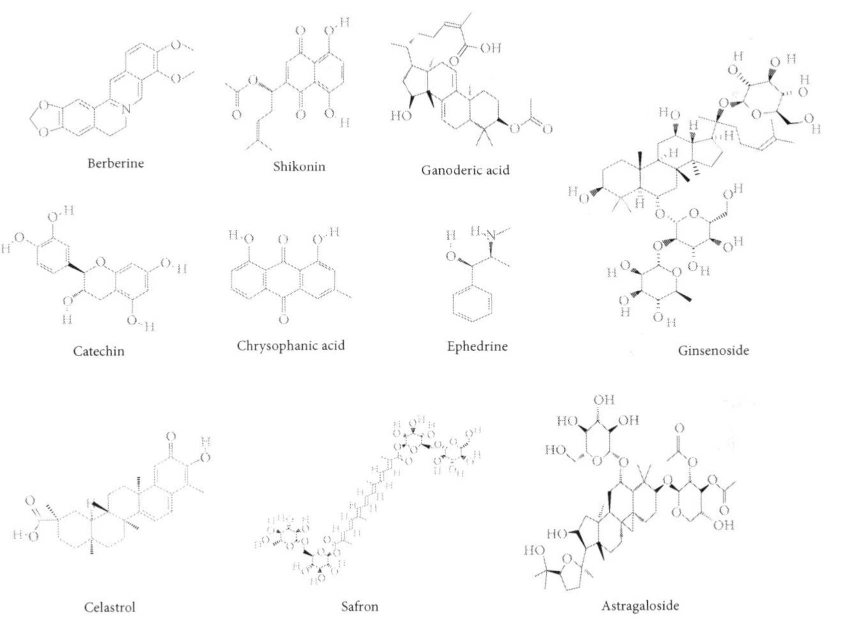

Figure 1: Molecular structures of the compounds described in this review.

Rhizoma Coptidis (Huang Lian)

Rhizoma coptidis is derived from the root of Coptis chinensis Franch., Coptis deltoidea C. Y. Cheng et Hsiao, orCoptis teeta Wall [55]. Its main components include alkaloids and lignans. Among the alkaloids, berberine is a main active component of Rhizoma coptidis [34]. The studies found that Rhizoma coptidis can reduce weight, lower lipids [56], reduce lipid synthesis [57], and inhibit adipogenesis [58]. Xie et al. [35] found that Rhizoma coptidis (RC) (200 mg/kg) and berberine (200 mg/kg) significantly lowered body and visceral adipose weight, reduced blood glucose and lipid levels, and decreased degradation of dietary polysaccharides in high-fat diet (HFD) mice. Both the ex vivo and in vitro trials confirmed that RC and berberine can regulate gut microbes to reduce weight. The antiobesity mechanisms of RC and berberine involve decreasing degradation of dietary polysaccharides, lowering caloric intake, and systemically activating fasting-induced adipose factor (FIAF) protein and expression of genes related to mitochondrial energy metabolism. Zhang et al. [59] found that when 3T3-L1 preadipocytes were cultured with various concentrations of berberine (0, 0.5, 1, 5, or 10 μM) for 7 days, berberine inhibited their differentiation. Significant inhibition of intracellular lipid accumulation was observed when 3T3-L1-derived adipocytes were exposed to berberine on days 3–5 and days 5–7, and this effect was marked at 5 μM berberine. The authors concluded that berberine suppresses adipocyte differentiation mainly by suppressing cAMP response element-binding protein (CREB) activity, which leads to a decrease in CCAAT/enhancer-binding protein beta- (C/EBPβ-) triggered transcriptional cascades.

Panax ginseng C. A. Mey (Ren Shen)

Ren Shen is derived from the dried root and rhizome of Panax ginseng C. A. Mey. (Araliaceae) [55]. Ginseng saponins and polysaccharides are the main active components of Panax ginseng C. A. Mey [60]. Ginseng saponins can be subdivided based on structure into Rb1, Rb2, Rc, Rd, Re, and Rl [36, 37]. Panax ginseng C. A. Mey can reduce body weight [61], attenuate fat accumulation [62], suppress lipid accumulation and reactive oxygen species (ROS) production [63], and improve insulin resistance [64]. Li et al. [65] found that administration of ginseng (0.5 g/kg diet) to HFD-induced obese mice for 15 weeks significantly decreased body fat mass gain, improved glucose tolerance and insulin sensitivity, and prevented hypertension. Koh et al. [63] investigated the treatment of 3T3-L1 cells with Rg1 (0, 25, 50, 100, and 200 μM). They observed that administration of 100 μM Rg1 for 24 h greatly reduced lipid accumulation and ROS production; treatment with 100 μM Rg1 in the early stages of 3T3-L1 differentiation (days 0–2) significantly decreased adipocyte formation. Rg1 reduces lipid accumulation and ROS production via the activation of C/EBP-homologous protein 10 (CHOP10), which attenuates fat accumulation and downregulates protein levels of NADPH oxidase 4 (NOX4). Lin et al. [66] found that when daily injections of 20 mg/kg Rb1 were administered to diet-induced obese mice for 3 weeks, body weight, food intake, blood glucose, and lipid levels decreased significantly. The ginsenoside, Rb1, may treat obesity by modifying the serum content and mRNA expressions of neuropeptide Y (NPY), NPY Y2 receptor, and peptide YY (PYY).

Radix Lithospermi (Zicao)

Radix Lithospermi is derived from the root of Arnebia euchroma (Royle) Johnst., Lithospermum erythrorhizonSieb. et Zucc, or Arnebia guttata Bunge [55]. Studies have shown that Radix Lithospermi can reduce weight, inhibit lipid accumulation, induce lipolysis, and regulate lipid metabolism. The main active ingredients of Radix Lithospermi are shikonin and acetylshikonin [38, 39, 67, 68]. Su et al. [40] found that intragastric administration of 100, 300, or 900 mg/kg acetylshikonin extract for 6 weeks in obese rats significantly decreased weight, serum free fatty acid (FFA), and serum triglyceride (TG) levels. Acetylshikonin is effective in the treatment of obesity by suppressing the expression of adipogenic differentiation transcription factors and adipocyte-specific proteins, and by increasing the activity of cAMP-dependent protein kinase (PKA) and phosphorylation of hormone-sensitive lipase (HSL). Su et al. [41] found that oral gavage of 540 mg/kg/day of acetylshikonin for 8 weeks in db/db mice significantly decreased body weight, food efficiency ratio, serum TG, and FFA levels. The mechanism of acetylshikonin activity in the treatment of obesity and nonalcoholic fatty liver disease involves the regulation of lipid metabolism and anti-inflammatory effects. Bettaieb et al. [42] found that administration of shikonin (2 mg/kg/day) to HFD mice for 5 days at an injected volume of 1% of their body weight could improve glucose tolerance and decrease body weight, adiposity, and hepatic dyslipidemia over 18 weeks. Shikonin acts by enhancing hepatic insulin signaling, increasing tyrosine phosphorylation of the insulin receptor, and enhancing downstream signaling.

Ephedra sinica Stapf. (Ma Huang)

The dried rhizome of Ephedra sinica Stapf. is used as the main ingredient of the herbal medicine, Ma Huang [69]. It has been used in recent years to treat obesity [43]. Other species that are used include Ephedra intermedia Schrenk et C. A. Mey. and Ephedra equisetina Bge [55]. Ephedra sinica Stapf. can modulate gut microbiota, reduce weight, and improve glucose intolerance. Song et al. [70] found that oral gavage of 5%Ephedra and 0.5% acarbose for 6 weeks in HFD-fed mice could reduce weight gain and epididymal fat accumulation, decrease fasting blood glucose, and improve lipid profiles and glucose intolerance. The main mechanism of Ephedra sinica's ability to reduce obesity and hyperglycemia involves increasing peroxisome proliferator-activated receptor alpha (PPAR-α) and adiponectin activity and reducing tumor necrosis factor-alpha (TNF-α) activity. The study published by Wang et al. [44] showed that administration of Ephedra sinica to HFD-induced obese rats by oral gavage over three weeks led to significant loss of body weight, epididymal fat, and perirenal fat, but no remarkable changes were observed in abdominal fat, liver weights, cecum weights, or food efficiency ratios.

Rheum palmatum L. (Da-Huang)

Da-Huang is derived from the dried root and rhizome of Rheum palmatum L., Rheum tanguticum Maxim. ex Balf, or Rheum officinale Baill [55]. Emodin and chrysophanic acid are the active compounds of Rheum palmatum L. [71]. Lim et al. [45] found that, following administration of chrysophanic acid (5 mg/kg/day) for 16 weeks to HFD mice, body weight, food intake, total cholesterol, low density lipoprotein (LDL) cholesterol, TG, and blood glucose decreased. In in vitro experiments, cells were cultured in medium containing chrysophanic acid for 48 h, and the results suggested that chrysophanic acid could suppresses lipid accumulation and downregulate adipogenic factors. Chrysophanic acid can ameliorate obesity by activating 5′-AMP-activated protein kinase alpha (the catalytic subunit of AMPK) to control the adipogenic and thermogenic pathway. Li et al. [46] found that administration of emodin (80 mg/kg/day) for 6 weeks to HFD-induced obese mice reduced body weights and fasting blood glucose levels, while improving insulin intolerance and serum and hepatic lipid levels. Emodin likely exerts its antiobesity effect by regulating the sterol regulatory element-binding protein (SREBP) pathway.

Green Tea (Lvcha)

Green tea, one of the most popular teas in China, contains tea polyphenols, catechins, caffeine, and amino acids; it is frequently used to ensure weight loss [47]. Green tea induces weight loss in a variety of ways, such as activating the nuclear factor erythroid-2-related-factor-2 (Nrf2) pathway [72], upregulation of neprilysin [73], prevention of gut dysbiosis [74], regulating metabolic balance in the body, inhibiting fat accumulation and cholesterol synthesis, and reducing abdominal fat. Choi et al. [48] found that administration of an HFD plus 0.25% (w/w) green tea extract for 12 weeks in diet-induced obesity (DIO) mice ameliorated obesity, hepatic steatosis, dyslipidemia, and insulin resistance. Green tea extract contributed to the regulation of systemic metabolic homeostasis via transcriptional responses to lipid, glucose, and amino acid metabolism. Zhu et al. [75] treated 3T3-L1 cells with catechins and caffeine in various concentrations and combinations for 8 or 12 days. Combination therapy with catechins and caffeine markedly reduced intracellular fat accumulation by regulating the gene and protein expression levels of lipid metabolism-related enzymes. Yamashita et al. [76] found that, following supplementation with green tea extract powder and eriodictyol for 8 weeks, body weight, food intake, cholesterol levels, and LDL levels were decreased, accompanied by the suppression of two kinds of cholesterol synthesis enzymes, 3-hydroxy-3-methylglutaryl-coenzyme A reductase (HMGCR), and 3-hydroxy-3-methylglutaryl-coenzyme A synthase (HMGCS).

Astragalus membranaceus (Fisch.) Bunge (Huang Qi or Radix Astragali)

Huang qi is derived from the dried root of Astragalus membranaceus (Fisch.) Bunge var. mongholicus orAstragalus membranaceus (Fisch.) Bunge [55]. The main active components of Astragalus membranaceus(Fisch.) Bunge are astragaloside, campanulin, ononin, kaempferol, and astragalus polysaccharides [77]. Xu et al. [49] found that when cells were incubated with isoastragaloside I (HQ1) and astragalosides II (HQ2), extracts of Radix Astragali, for 72 h, insulin resistance, and glucose intolerance were improved. Oral gavage with HQ1 and HQ2 (50 mg of each compound/kg body weight, twice a day) for 6 weeks in db/db mice increased serum levels of total adiponectin, possibly via activation of AMPK. The study published by Hoo et al. [78] suggested that daily oral gavage with Radix Astragali (2 g/kg/day) in db/db diabetic mice for 12 weeks reduces body weight and food intake and alleviates glucose intolerance/insulin resistance. The main mechanism may be the suppression of inflammatory pathways.

Carthamus tinctorius L. (Hong Hua)

Carthamus tinctorius L. is derived from the dried flower of Crocus sativus L. The main active component ofCarthamus tinctorius L. is saffron [50]. The study published by Zhu et al. [52] showed that HFD-induced obese male ICR mice, intraperitoneally injected with safflower yellow (120 mg/kg) daily for eight weeks, had significant reductions in body fat mass, fasting blood glucose, and improvements in insulin sensitivity. A possible mechanism is the promotion of the browning of subcutaneous white adipose tissue (WAT) and activating the insulin receptor substrate 1/Akt/glycogen synthase kinase 3β pathway in visceral WAT. Mashmoul et al. [79] used saffron (dried stigma of Crocus sativus L. flowers) to treat obesity-related fatty liver. They were administered saffron extract and crocin at concentrations of 40 and 80 mg/kg/day for 8 weeks in HFD-induced obese rats. Levels of liver enzymes, relative liver weights, and food intake were decreased. Saffron had a curative effect in the treatment of obesity-related fatty liver disease, but more definitive evidence of the protective effects of saffron and crocin needs to be found.

Ganoderma lucidum (Leyss. ex Fr.) Karst. (Lingzhi)

Lingzhi is derived from the dried fruiting body of Ganoderma lucidum (Leyss. ex Fr.) [55]. Chang et al. [80] found that daily treatments for 2 months with 100 μL of the water extract of Ganoderma lucidum mycelium at 2, 4, or 8% (w/v) by intragastric gavage in obese mice decreased weight gain and fat accumulation and decreased proinflammatory cytokine expression in the liver and adipose tissues in a dose-dependent manner. The 8% water extract of Ganoderma lucidum mycelium was the most effective treatment for modulating gut microbiota. The results indicate that Ganoderma lucidum reduces obesity in mice by modulating the composition of the gut microbiota, reducing endotoxemia, and preventing insulin resistance. Thyagarajan-Sahu et al. [81] found that treatment of 3T3-L1 preadipocytes with ReishiMax (RM, containing Ganoderma lucidum) (0–300 μg/mL) for 9 days decreased lipid accumulation, triglyceride uptake, and glycerol accumulation in a concentration-dependent manner. RM can control adipocyte differentiation and glucose uptake, possibly via suppressed expression of the adipogenic transcription factor, PPAR-γ, and enzymes and proteins responsible for lipid synthesis.

Tripterygium wilfordii Hook. f (Lei Gong Teng or Thunder God Vine)

Lei Gong Teng is derived from the dried roots, leaves, and flowers of Tripterygium wilfordii Hook. f [55]. Celastrol is the main active ingredient of Tripterygium wilfordii Hook. f [82]. Liu et al. [53] administered Celastrol (100 μg/kg) intraperitoneally for three weeks to HFD-induced obese mice and found that Celastrol suppressed food intake, improved energy expenditure, and leads up to 45% weight loss in hyperleptinemic diet-induced obese mice by increasing leptin sensitivity. Following treatment with vehicle or Celastrol (100 μg/kg) (daily, intraperitoneal) in db/db or ob/ob mice, food intake slightly decreased during the first week, and body weight, lean mass, and fat percentage were not affected by Celastrol treatment. Celastrol is a leptin sensitizer, and its main mechanism of weight reduction is relief of endoplasmic reticulum (ER) stress and increased leptin sensitivity. In another study [83], Hu et al. identified Nur77 as a critical intracellular target of Celastrol, which induces apoptosis by targeting mitochondria. Hu et al. used their findings to develop a safe and effective drug to reduce weight. Thunder god vine should be used cautiously because of its complex composition [54] and potential adverse reactions [84].

CONCLUSIONS AND PERSPECTIVES

The effect and the relevant mechanisms behind how herbal medicine work as an antiobesity treatment are still controversial. During the past decade, much recent progress has been made in the study of weight loss therapy with herbal medicine. Clinical investigations of herbal medicine have been shown to be effective in the treatment of obesity, and animal experiments have begun to reveal the potential mechanisms of the various herbal medicine. However, there are some limitations as follows: (1) Obesity is associated with oxidative stress, but there have been fewer reports in this area. Flos carthami has been shown to be effective against oxidative stress and further study of oxidative stress and weight loss using safflower is warranted. (2) Some herbal treatments also show some toxicity and should be used with caution. For example, the drug composition of thunder god vine is complex, and when it is used to treat obesity, liver and kidney function should be closely monitored. There are many herbal medicines that have adverse effects if used on long-term or at the incorrect dosages, so the long-term application of herbal medicine for obesity should focus on the safety evaluation; for example, in one case [18], a skin rash was reported in the XJXGF formula group, but the rash was transient and disappeared without treatment. (3) Clinical reports indicate that herbal medicines for obesity produce few adverse reactions, and their level of safety is acceptable. However, some cases of adverse reactions have been reported, such as a case of sudden death due to the use of green tea. Therefore, the use of traditional Chinese medicines should be regulated. (4) The drug composition of herbal medicine is complex, making it difficult to determine the mechanism(s) of action, unlike in western medicine. There was also a case report [85] of a 19-year-old obese man (120 kg) who drank large amounts of green tea (15 cups per day) with a strict diet regimen, over 2 months; he lost 30 kg of body weight. However, after his usual exercise, he died of left ventricular fibrillation. His most prevalent symptoms were gastrointestinal problems, such as dyspepsia, epigastric pain, and nausea, as well as headache. Only a small number of the studies included herein have reported that the use of herbal medicine preparations caused adverse reactions. The safety of long-term use of herbal medicine needs to be further explored.

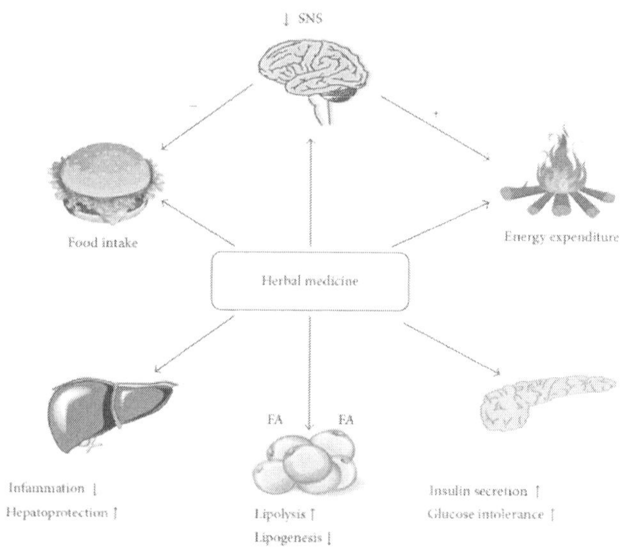

Figure 2: Potential mechanisms of herbal medicine for obesity. Herbal medicine such asCelastrol, its main mechanism of weight reduction is inhibiting endoplasmic reticulum (ER) stress and increased leptin sensitivity. Ganoderma lucidum mycelium decreased proinflammatory cytokine expression in the liver and adipose tissues in a dose-dependent manner; Acetylshikonin covers the treatment of obesity which involves the regulation of lipid metabolism and anti-inflammatory effects and decreased serum free fatty acid. Radix Astragali reduces body weight and food intake, and alleviates glucose intolerance/insulin resistance. SNS: sympathetic nervous system; FA: fatty acid.

Use of herbal medicine to treat obesity is currently garnering much attention. Only a small number of the active ingredients available in herbs have been identified, and if the composition of the herbs is more and more identified in the future, the target and definite mechanism of action can be determined. As mentioned above, herbal medicine has some beneficial effects on the treatment of patients with obesity and has fewer adverse effects than chemical agents; potential mechanisms of herbal medicine for obesity were presented in Figure 2. Extensive preclinical and clinical researches [86] have highlighted the pharmaceutical uses of herbal medicine as antidiabetic, antihyperlipidemic, antiobesity, anti-inflammatory, and antioxidant. In clinical practice, herbal medicines are usually used in a compound form. With the development of modern pharmacological science, it is easier to identify the active agents in herbal medicine compounds, facilitating scientific study of their effectiveness. In addition, more and more clinical trials and a standardized procedure of herbal medicine producing are needed to confirm the safety and antiobesity effect of herbal medicine and finally prevent/reduce obesity by herbal medicine consumption in human.

AUTHORS' CONTRIBUTIONS

Rui Gao and Yue Liu contributed to the topic conception, manuscript revision, and the decision to submit for publication and are the co-corresponding authors. Yanfei Liu and Mingyue Sun put on the references collection, references analysis, and writing of the manuscript together, contributed equally to this work, and are the co-first authors. Hezhi Yao contributed to references analysis and helped in revising manuscript.

ACKNOWLEDGMENTS

The authors gratefully acknowledge the financial support from Special Science and Technology Research Program of Traditional Chinese Medicine of State Administration of Traditional Chinese Medicine (2016ZX10) of Professor Rui Gao, National Natural Science Foundation of China (Grant no. 81403266), and Beijing NOVA Program (no. Z171100001117027) of Dr. Yue Liu.

REFERENCES

1. Y. C. Wang, K. McPherson, T. Marsh, S. L. Gortmaker, and M. Brown, "Health and economic burden of the projected obesity trends in the USA and the UK," The Lancet, vol. 378, no. 9793, pp. 815–825, 2011.
2. https://www.aace.com/sites/all/files/Obesity_Guidelines_Algorithm_slides_FINAL_2016.pdf.
3. The GBD 2015 Obesity Collaborators, "Health Effects of Overweight and Obesity in 195 Countries over 25 Years," New England Journal of Medicine, vol. 377, no. 1, pp. 13–27, 2017.
4. A. Obata, S. Okauchi, T. Kimura et al., "Advanced breast cancer in a relatively young man with severe obesity and type 2 diabetes mellitus," Journal of Diabetes Investigation, vol. 8, no. 3, pp. 395-396, 2017.
5. M. Ng, T. Fleming, M. Robinson, and et al, "Global, regional, and national prevalence of overweight and obesity in children and adults during 1980–2013: A systematic analysis for the Global Burden of Disease Study 2013," The Lancet, vol. 384, no. 9945, pp. 766–781, 2014.
6. L. Landsberg, L. J. Aronne, L. J. Beilin et al., "Obesity-related hypertension: Pathogenesis, cardiovascular risk, and treatment—a position paper of the obesity society and the American society of hypertension," Obesity, vol. 21, no. 1, pp. 8–24, 2013.
7. C. Arroyo-Johnson and K. D. Mincey, "Obesity Epidemiology Worldwide," Gastroenterology Clinics of North America, vol. 45, no. 4, pp. 571–579, 2016.
8. P. L. Lefèbvre and A. J. Scheen, "Obesity: Causes and new treatments," Experimental and Clinical Endocrinology and Diabetes, vol. 109, no. 2, pp. S215–S224, 2001.
9. C. M. Apovian, L. J. Aronne, D. H. Bessesen et al., "Pharmacological management of obesity: an endocrine society clinical practice guideline," Journal of Clinical Endocrinology and Metabolism, vol. 100, no. 2, pp. 342–362, 2015.
10. M. O. Dietrich and T. L. Horvath, "Limitations in anti-obesity drug development: the critical role of hunger-promoting neurons," Nature Reviews Drug Discovery, vol. 11, no. 9, pp. 675–691, 2012.
11. E. Mead, G. Atkinson, B. Richter et al., "Drug interventions for the treatment of obesity in children and adolescents," Cochrane Database

of Systematic Reviews, vol. 29, no. 11, Article ID CD012436, 2016.

12. C. Van Der Schoor, H. M. Oberholzer, M. J. Bester, and M.-J. Van Rooy, "The effect of sibutramine, a serotonin-norepinephrine reuptake inhibitor, on platelets and fibrin networks of male Sprague-Dawley rats: A descriptive study," Ultrastructural Pathology, vol. 38, no. 6, pp. 399–405, 2014.
13. J. J. Kim, M. E. Tarnoff, and S. A. Shikora, "Surgical Treatment for Extreme Obesity: Evolution of a Rapidly Growing Field," Nutrition in Clinical Practice, vol. 18, no. 2, pp. 109–123, 2003.
14. S. Ikramuddin, J. Korner, W.-J. Lee et al., "Durability of addition of Roux-en-Y Gastric Bypass to lifestyle intervention and medical management in achieving primary treatment goals for uncontrolled type 2 diabetes in mild to moderate obesity: A randomized control trial," Diabetes Care, vol. 39, no. 9, pp. 1510–1518, 2016.
15. J. Y. Park and Y. J. Kim, "Laparoscopic Roux-en-Y gastric bypass in obese Korean patients: efficacy and potential adverse events," Surgery Today, vol. 46, no. 3, pp. 348–355, 2016.
16. J. E. Lambert, J. A. Parnell, J. M. Tunnicliffe, J. Han, T. Sturzenegger, and R. A. Reimer, "Consuming yellow pea fiber reduces voluntary energy intake and body fat in overweight/obese adults in a 12-week randomized controlled trial," Clinical Nutrition, 2015.
17. K. Azushima, K. Tamura, S. Haku et al., "Effects of the oriental herbal medicine Bofu-tsusho-san in obesity hypertension: A multicenter, randomized, parallel-group controlled trial (ATH-D-14-01021. R2)," Atherosclerosis, vol. 240, no. 1, pp. 297–304, 2015.
18. Q. Zhou, B. Chang, X.-Y. Chen et al., "Chinese herbal medicine for obesity: A randomized, double-blinded, multicenter, prospective trial," American Journal of Chinese Medicine, vol. 42, no. 6, pp. 1345–1356, 2014.
19. G. B. Lenon, K. X. Li, Y.-H. Chang et al., "Efficacy and safety of a Chinese herbal medicine formula (RCM-104) in the management of simple obesity: A randomized, placebo-controlled clinical trial," Evidence-based Complementary and Alternative Medicine, vol. 2012, Article ID 435702, 2012.
20. B. Ke, L. Shi, J.-J. Zhang, D.-S. Chen, J. Meng, and J. Qin, "Protective effects of Modified Linggui Zhugan Decoction combined with short-term very low calorie diets on cardiovascular risk factors in obese

patients with impaired glucose tolerance," Journal of Traditional Chinese Medicine, vol. 32, no. 2, pp. 193–198, 2012.

21. S.-L. Chu, H. Fu, J.-X. Yang et al., "A randomized double-blind placebo-controlled study of Pu'er tea extract on the regulation of metabolic syndrome," Chinese Journal of Integrative Medicine, vol. 17, no. 7, pp. 492–498, 2011.
22. Z. Li, R. Song, C. Nguyen et al., "Pistachio nuts reduce triglycerides and body weight by comparison to refined carbohydrate snack in obese subjects on a 12-week weight loss program," Journal of the American College of Nutrition, vol. 29, no. 3, pp. 198–203, 2010.
23. M. Abidov, Z. Ramazanov, R. Seifulla, and S. Grachev, "The effects of Xanthigen in the weight management of obese premenopausal women with non-alcoholic fatty liver disease and normal liver fat," Diabetes, Obesity and Metabolism, vol. 12, no. 1, pp. 72–81, 2010.
24. B. Gout, C. Bourges, and S. Paineau-Dubreuil, "Satiereal, a Crocus sativus L extract, reduces snacking and increases satiety in a randomized placebo-controlled study of mildly overweight, healthy women," Nutrition Research, vol. 30, no. 5, pp. 305–313, 2010.
25. E. A. Datau, Wardhana, E. E. Surachmanto, K. Pandelaki, J. A. Langi, and Fias, "Efficacy of Nigella sativa on serum free testosterone and metabolic disturbances in central obese male," Acta Medica Indonesiana, vol. 42, no. 3, pp. 130–134, 2010.
26. F. Di Pierro, A. B. Menghi, A. Barreca, M. Lucarelli, and A. Calandrelli, "GreenSelect phytosome as an adjunct to a low-calorie diet for treatment of obesity: A clinical trial," Alternative Medicine Review, vol. 14, no. 2, pp. 154–160, 2009.
27. H. Wang, Y. Wen, Y. Du et al., "Effects of catechin enriched green tea on body composition," Obesity, vol. 18, no. 4, pp. 773–779, 2010.
28. R.-R. He, L. Chen, B.-H. Lin, Y. Matsui, X.-S. Yao, and H. Kurihara, "Beneficial effects of oolong tea consumption on diet-induced overweight and obese subjects," Chinese Journal of Integrative Medicine, vol. 15, no. 1, pp. 34–41, 2009.
29. J. L. Ngondi, B. C. Etoundi, C. B. Nyangono, C. M. F. Mbofung, and J. E. Oben, "IGOB131, a novel seed extract of the West African plant Irvingia gabonensis, significantly reduces body weight and improves metabolic parameters in overweight humans in a randomized double-blind placebo controlled investigation," Lipids in Health and Disease,

vol. 8, article no. 7, 2009.

30. J. E. Oben, J. L. Ngondi, C. N. Momo, G. A. Agbor, and C. S. M. Sobgui, "The use of a Cissus quadrangularis/Irvingia gabonensis combination in the management of weight loss: A double-blind placebo-controlled study," Lipids in Health and Disease, vol. 7, article no. 12, 2008.
31. C. Roongpisuthipong, R. Kantawan, and W. Roongpisuthipong, "Reduction of adipose tissue and body weight: effect of water soluble calcium hydroxycitrate in Garcinia atroviridis on the short term treatment of obese women in Thailand," Asia Pacific Journal of Clinical Nutrition, vol. 16, no. 1, pp. 25–29, 2007.
32. R. Kuriyan, T. Raj, S. K. Srinivas, M. Vaz, R. Rajendran, and A. V. Kurpad, "Effect of Caralluma Fimbriata extract on appetite, food intake and anthropometry in adult Indian men and women," Appetite, vol. 48, no. 3, pp. 338–344, 2007.
33. J. E. Oben, D. M. Enyegue, G. I. Fomekong, Y. B. Soukontoua, and G. A. Agbor, "The effect of Cissus quadrangularis (CQR-300) and a Cissus formulation (CORE) on obesity and obesity-induced oxidative stress," Lipids in Health and Disease, vol. 6, article no. 4, 2007.
34. W.-J. Ni, H.-H. Ding, and L.-Q. Tang, "Berberine as a promising anti-diabetic nephropathy drug: An analysis of its effects and mechanisms," European Journal of Pharmacology, vol. 760, pp. 103–112, 2015.
35. W. Xie, D. Gu, J. Li, K. Cui, and Y. Zhang, "Effects and action mechanisms of berberine and rhizoma coptidis on gut microbes and obesity in high-fat diet-fed C57BL/6J mice," PLoS ONE, vol. 6, no. 9, Article ID e24520, 2011.
36. L. P. Christensen, "Chapter 1 ginsenosides: chemistry, biosynthesis, analysis, and potential health effects," Advances in Food and Nutrition Research, vol. 55, pp. 1–99, 2008.
37. S. A. Palaniyandi, J. W. Suh, and S. H. Yang, "Preparation of ginseng extract with enhanced levels of ginsenosides Rg1 and Rb1 using high hydrostatic pressure and polysaccharide hydrolases," Pharmacognosy Magazine, vol. 13, 1, no. 49, pp. 142–147, 2017.
38. X. Zhang, J. Cui, Q. Meng, S. Li, W. Zhou, and S. Xiao, "Advance in anti-tumor mechanisms of shikonin, alkannin and their derivatives," Mini-Reviews in Medicinal Chemistry, vol. 17, no. 18, 2017.
39. X. Chen, L. Yang, J. J. Oppenheim, and O. M. Z. Howard, "Cellular

pharmacology studies of shikonin derivatives," Phytotherapy Research, vol. 16, no. 3, pp. 199–209, 2002.

40. M. Su, W. Huang, and B. Zhu, "Acetylshikonin from Zicao prevents obesity in rats on a high-fat diet by inhibiting lipid accumulation and inducing lipolysis," PLoS ONE, vol. 11, no. 1, Article ID e0146884, 2016.
41. M.-L. Su, Y. He, Q.-S. Li, and B.-H. Zhu, "Efficacy of acetylshikonin in preventing obesity and hepatic steatosis in db/db mice," Molecules, vol. 21, no. 8, article 976, 2016.
42. A. Bettaieb, E. Hosein, S. Chahed et al., "Decreased adiposity and enhanced glucose tolerance in shikonin treated mice," Obesity, vol. 24, no. 5, pp. 2269–2277, 2016.
43. E. A. Abourashed, A. T. El-Alfy, I. A. Khan, and L. Walker, "Ephedra in perspective—a current review," Phytotherapy Research, vol. 17, no. 7, pp. 703–712, 2003.
44. J.-H. Wang, B.-S. Kim, K. Han, and H. Kim, "Ephedra-treated donor-derived gut microbiota transplantation ameliorates high fat diet-induced obesity in rats," International Journal of Environmental Research and Public Health, vol. 14, aricle 555, no. 6, 2017.
45. H. Lim, J. Park, H.-L. Kim et al., "Chrysophanic acid suppresses adipogenesis and induces thermogenesis by activating AMP-activated protein kinase alpha in vivo and in vitro," Frontiers in Pharmacology, vol. 4, article 476, no. 7, 2016.
46. J. Li, L. Ding, B. Song et al., "Emodin improves lipid and glucose metabolism in high fat diet-induced obese mice through regulating SREBP pathway," European Journal of Pharmacology, vol. 770, pp. 99–109, 2016.
47. D. Türközü and N. A. Tek, "A minireview of effects of green tea on energy expenditure," Critical Reviews in Food Science and Nutrition, vol. 57, no. 2, pp. 254–258, 2017.
48. J.-Y. Choi, Y. J. Kim, R. Ryu, S.-J. Cho, E.-Y. Kwon, and M.-S. Choi, "Effect of green tea extract on systemic metabolic homeostasis in diet-induced obese mice determined via RNA-seq transcriptome profiles," Nutrients, vol. 8, no. 10, article no. 640, 2016.
49. A. Xu, H. Wang, R. L. C. Hoo et al., "Selective elevation of adiponectin production by the natural compounds derived from a medicinal herb alleviates insulin resistance and glucose intolerance in obese

mice," Endocrinology, vol. 150, no. 2, pp. 625–633, 2009.

50. E. Christodoulou, N. P. Kadoglou, N. Kostomitsopoulos, and G. Valsami, "Saffron: A natural product with potential pharmaceutical applications," Journal of Pharmacy and Pharmacology, vol. 67, no. 12, pp. 1634–1649, 2015.
51. M. Mashmoul, A. Azlan, N. Mohtarrudin et al., "Protective effects of saffron extract and crocin supplementation on fatty liver tissue of high-fat diet-induced obese rats," BMC Complementary and Alternative Medicine, vol. 16, no. 1, article 401, 2016.
52. H. Zhu, X. Wang, H. Pan et al., "The mechanism by which safflower yellow decreases body fat mass and improves insulin sensitivity in HFD-induced obese mice," Frontiers in Pharmacology, vol. 7, no. MAY, article 127, 2016.
53. J. Liu, J. Lee, M. Salazar Hernandez, R. Mazitschek, and U. Ozcan, "Treatment of obesity with celastrol," Cell, vol. 161, no. 5, pp. 999–1011, 2015.
54. S. K.-Y. Law, M. P. Simmons, N. Techen et al., "Molecular analyses of the Chinese herb Leigongteng (Tripterygium wilfordii Hook.f.)," Phytochemistry, vol. 72, no. 1, pp. 21–26, 2011.
55. Chinese Pharmacopoeia Commission, Pharmacopoeia of the People's Republic of China, Chinese Medical Science Press, Beijing, China, 2010.
56. Z.-Y. Zou, Y.-R. Hu, H. Ma et al., "Coptisine attenuates obesity-related inflammation through LPS/TLR-4-mediated signaling pathway in Syrian golden hamsters," Fitoterapia, vol. 105, pp. 139–146, 2015.
57. W.-L. Zhang, L. Zhu, and J.-G. Jiang, "Active ingredients from natural botanicals in the treatment of obesity," Obesity Reviews, vol. 15, no. 12, pp. 957–967, 2014.
58. J. S. Choi, J.-H. Kim, M. Y. Ali, B.-S. Min, G.-D. Kim, and H. A. Jung, "Coptis chinensis alkaloids exert anti-adipogenic activity on 3T3-L1 adipocytes by downregulating C/EBP-α and PPAR-γ," Fitoterapia, vol. 98, pp. 199–208, 2014.
59. J. Zhang, H. Tang, R. Deng et al., "Berberine suppresses adipocyte differentiation via decreasing CREB transcriptional activity," PLoS ONE, vol. 10, no. 4, Article ID e0125667, 2015.
60. Y. Li, T. J. Zhang, S. X. Liu, R. R. Wang, and J. Luo, "Research on chemical compositions and pharmacology role of Gingeng," Chinese

Traditional and Herbal Drugs, vol. 40, no. 1, pp. 164–166, 2009.

61. M. C. Kho, Y. J. Lee, J. H. Park et al., "Fermented red ginseng potentiates improvement of metabolic dysfunction in metabolic syndrome rat models," Nutrients, vol. 8, no. 6, article 369, 2016.
62. N. Karu, R. Reifen, and Z. Kerem, "Weight gain reduction in mice fed Panax ginseng saponin, a pancreatic lipase inhibitor," Journal of Agricultural and Food Chemistry, vol. 55, no. 8, pp. 2824–2828, 2007.
63. E.-J. Koh, K.-J. Kim, J. Choi, H. J. Jeon, M.-J. Seo, and B.-Y. Lee, "Ginsenoside Rg1 suppresses early stage of adipocyte development via activation of C/EBP homologous protein-10 in 3T3-L1 and attenuates fat accumulation in high fat diet-induced obese zebrafish," Journal of Ginseng Research, 2015.
64. J.-M. Cheon, D.-I. Kim, and K.-S. Kim, "Insulin sensitivity improvement of fermented Korean red ginseng (Panax ginseng) mediated by insulin resistance hallmarks in old-aged ob/ob mice," Journal of Ginseng Research, vol. 39, no. 4, pp. 331–337, 2015.
65. X. Li, J. Luo, P. V. A. Babu et al., "Dietary supplementation of chinese ginseng prevents obesity and metabolic syndrome in high-fat diet-fed mice," Journal of Medicinal Food, vol. 17, no. 12, pp. 1287–1297, 2014.
66. N. Lin, D. Cai, D. Jin, Y. Chen, and J. Shi, "Ginseng panaxoside Rb1 reduces body weight in diet-induced obese mice," Cell Biochemistry and Biophysics, vol. 68, no. 1, pp. 189–194, 2014.
67. S.-H. Park, N. M. Phuc, J. Lee et al., "Identification of acetylshikonin as the novel CYP2J2 inhibitor with anti-cancer activity in HepG2 cells," Phytomedicine, vol. 24, pp. 134–140, 2017.
68. Y. He, Q. Li, M. Su, W. Huang, and B. Zhu, "Acetylshikonin from Zicao exerts antifertility effects at high dose in rats by suppressing the secretion of GTH," Biochemical and Biophysical Research Communications, vol. 476, no. 4, pp. 560–565, 2016.
69. R. Gul, S. U. Jan, S. Faridullah, S. Sherani, and N. Jahan, "Preliminary Phytochemical Screening, Quantitative Analysis of Alkaloids, and Antioxidant Activity of Crude Plant Extracts from Ephedra intermedia Indigenous to Balochistan," Scientific World Journal, vol. 2017, Article ID 5873648, 2017.
70. M.-K. Song, J.-Y. Um, H.-J. Jang, and B.-C. Lee, "Beneficial effect of dietary Ephedra sinica on obesity and glucose intolerance in high-fat

diet-fed mice," Experimental and Therapeutic Medicine, vol. 3, no. 4, pp. 707–712, 2012.

71. L. Liu, L. Fan, H. Chen, X. Chen, and Z. Hu, "Separation and determination of four active anthraquinones in Chinese herbal preparations by flow injection-capillary electrophoresis," Electrophoresis, vol. 26, no. 15, pp. 2999–3006, 2005.
72. C. Sampath, M. R. Rashid, S. Sang, and M. Ahmedna, "Green tea epigallocatechin 3-gallate alleviates hyperglycemia and reduces advanced glycation end products via nrf2 pathway in mice with high fat diet-induced obesity," Biomedicine and Pharmacotherapy, vol. 87, pp. 73–81, 2017.
73. M. Muenzner, N. Tappenbeck, F. Gembardt et al., "Green tea reduces body fat via upregulation of neprilysin," International Journal of Obesity, vol. 40, no. 12, pp. 1850–1855, 2016.
74. M. Cheng, X. Zhang, Y. Miao, J. Cao, Z. Wu, and P. Weng, "The modulatory effect of (-)-epigallocatechin 3-O-(3-O-methyl) gallate (EGCG3"Me) on intestinal microbiota of high fat diet-induced obesity mice model," Food Research International, vol. 92, pp. 9–16, 2017.
75. X. Zhu, L. Yang, F. Xu, L. Lin, and G. Zheng, "Combination therapy with catechins and caffeine inhibits fat accumulation in 3T3-L1 cells," Experimental and Therapeutic Medicine, vol. 13, no. 2, pp. 688–694, 2017.
76. M. Yamashita, M. Kumazoe, Y. Nakamura et al., "The combination of green tea extract and eriodictyol inhibited high-fat/high-sucrose diet-induced cholesterol upregulation is accompanied by suppression of cholesterol synthesis enzymes," Journal of Nutritional Science and Vitaminology, vol. 62, no. 4, pp. 249–256, 2016.
77. H. Li, F. Xu, P. Yang et al., "Systematic screening and characterization of prototype constituents and metabolites of total astragalosides using HPLC-ESI-IT-TOF-MSn after oral administration to rats," Journal of Pharmaceutical and Biomedical Analysis, vol. 142, pp. 102–112, 2017.
78. R. L. C. Hoo, J. Y. L. Wong, C. F. Qiao, A. Xu, H. X. Xu, and K. S. L. Lam, "The effective fraction isolated from Radix Astragali alleviates glucose intolerance, insulin resistance and hypertriglyceridemia in db/db diabetic mice through its anti-inflammatory activity," Nutrition and Metabolism, vol. 7, article 67, 2010.
79. M. Mashmoul, A. Azlan, N. Mohtarrudin et al., "Protective effects of

saffron extract and crocin supplementation on fatty liver tissue of high-fat diet-induced obese rats," BMC Complementary and Alternative Medicine, vol. 16, article 401, no. 1, 2016.

80. C. Chang, C. Lin, C. Lu et al., "Ganoderma lucidum reduces obesity in mice by modulating the composition of the gut microbiota," Nature Communications, vol. 23, article 7489, no. 6, 2015.
81. A. Thyagarajan-Sahu, B. Lane, and D. Sliva, "ReishiMax, mushroom based dietary supplement, inhibits adipocyte differentiation, stimulates glucose uptake and activates AMPK," BMC Complementary and Alternative Medicine, vol. 11, article 74, 2011.
82. M. Li, X. Liu, Y. He et al., "Celastrol attenuates angiotensin II mediated human umbilical vein endothelial cells damage through activation of Nrf2/ERK1/2/Nox2 signal pathway," European Journal of Pharmacology, vol. 797, pp. 124–133, 2017.
83. M. Hu, Q. Luo, G. Alitongbieke et al., "Celastrol-induced Nur77 interaction with TRAF2 alleviates inflammation by promoting mitochondrial ubiquitination and autophagy," Molecular Cell, vol. 66, no. 1, pp. 141–153, 2017.
84. S. Zhao, F. Otieno, A. Akpan, and R. J. Moots, "Complementary and Alternative Medicine Use in Rheumatoid Arthritis: Considerations for the Pharmacological Management of Elderly Patients," Drugs and Aging, vol. 34, no. 4, pp. 255–264, 2017.
85. J. Najafian, M. Abdar-Esfahani, M. Arab-Momeni, and A. Akhavan-Tabib, "Safety of herbal medicine in treatment of weight loss," ARYA Atherosclerosis, vol. 10, no. 1, pp. 55–58, 2014.
86. J. Martel, D. M. Ojcius, C.-J. Chang et al., "Anti-obesogenic and antidiabetic effects of plants and mushrooms," Nature Reviews Endocrinology, vol. 13, no. 3, pp. 149–160, 2017.

CHAPTER 9

ECTOPIC FAT AND INSULIN RESISTANCE: PATHOPHYSIOLOGY AND EFFECT OF DIET AND LIFESTYLE INTERVENTIONS

M. Snel[1] , J. T. Jonker[1] , J. Schoones[2] , H. Lamb[3] , A. de Roos[3] , H. Pijl[1] , J. W. A. Smit[1] , A. E. Meinders[1] , and I. M. Jazet[1]

[1]Departments of Endocrinology & Metabolism and General Internal Medicine, Leiden University Medical Center, 2300 RC Leiden, The Netherlands

[2]Walaeus Library, Leiden University Medical Center, 2300 RC Leiden, The Netherlands

[3]Department of Radiology, Leiden University Medical Center, 2300 RC Leiden, The Netherlands

ABSTRACT

The storage of triglyceride (TG) droplets in nonadipose tissues is called ectopic fat storage. Ectopic fat is associated with insulin resistance and type 2 diabetes mellitus (T2DM). Not the triglycerides per se but the accumulation of intermediates of lipid metabolism in organs, such as the liver, skeletal

Citation: M. Snel, J. T. Jonker, J. Schoones, H. Lamb, A. de Roos, H. Pijl, J. W. A. Smit, A. E. Meinders, I. M. Jazet. Ectopic Fat and Insulin Resistance: Pathophysiology and Effect of Diet and Lifestyle Interventions, Hindawi Publishing Corporation International Journal of Endocrinology Volume 2012, Article ID 983814, 18 pages doi:10.1155/2012/983814

muscle, and heart seem to disrupt metabolic processes and impair organ function. We describe the mechanisms of ectopic fat depositions in the liver, skeletal muscle, and in and around the heart and the consequences for each organs function. In addition, we systematically reviewed the literature for the effects of diet-induced weight loss and exercise on ectopic fat depositions.

INTRODUCTION

The amount of people with obesity has increased dramatically over the past decades to an estimated number of 400 million adults worldwide with a projected 700 million in 2015 [1]. Obesity predisposes to the development of insulin resistance, type 2 diabetes mellitus (T2DM), and cardiovascular disease (CVD) [2–7]. However, about 30% of obese men and women are metabolically healthy [8], that is, do not have hypertension, dyslipidemia, or disturbances in glucose metabolism. Vice versa, these metabolic abnormalities occur in 20–30% of normal weight people.

Adipose tissue consists of adipocytes and the so-called stromal-vascular fraction that encompasses blood vessels and stroma with macrophages. Adipose tissue has the unique capacity to store large amounts of energy in the form of triglycerides (TG). For a long time, it has been presumed that this was the only function of adipose tissue. However, adipose tissue (i.e., adipocytes and the stromal vascular fraction) acts as an endocrine organ by secreting various hormones and cytokines (also referred to as adipokines) with effects on glucose and lipid metabolism and energy homeostasis [9]. It now appears that in those obese subjects that develop insulin resistance, adipose tissue dysfunction plays a role. Adipose tissue dysfunction is characterized among others by large adipocytes and secretion of adipokines with a proinflammatory profile ultimately leading to (among others) ectopic fat deposition [10]. Ectopic fat is defined as storage of TG in tissues other than adipose tissue, that normally contain only small amounts of fat, such as the liver, skeletal muscle, heart, and pancreas. Ectopic fat can interfere with cellular functions and hence organ function and is associated with insulin resistance.

The aim of this paper is to discuss the pathophysiology of adipose tissue dysfunction and ectopic fat deposition. We will elaborate on the effect of ectopic fat deposition on the cellular level as well as on the organ level for those organs involved in the insulin-resistant pathogenesis of T2DM (liver and skeletal muscle cell) and on the heart (because of the strongly increased risk for coronary heart disease in T2DM). Since there is uncertainty to date as to whether pancreatic steatosis is causally related to beta-cell dysfunction [11–13], and because it is not subject of investigation in our own studies, we will not discuss fat accumulation in the pancreas. Finally, because diet and exercise are the mainstays of therapy for obesity and T2DM, we systematically reviewed the literature for the effect of diet and or exercise on diminishing ectopic fat stores and restoring insulin sensitivity in obesity and T2DM.

METHODS SYSTEMATIC REVIEW

The following databases were searched: PubMed (1949 to January 2012), EMBASE (OVID-version, 1980 to January 2012), Web of Science (1945 to January 2012), and Cochrane Library (1990 to January 2012). The search strategy consisted of the AND combination of three main concepts:(1)type 2 diabetes mellitus, Obesity, or insulin resistance;(2)weight loss, diet, or exercise;(3)ectopic fat.

For these three concepts, all relevant keyword variations were used. References were limited to human studies, adults, written in English or Dutch. In addition, only studies that used techniques that can quantify the amount of lipid accumulation and measured insulin sensitivity were included. Studies using surrogate markers for lipid accumulation (e.g., alanine aminotransferase (ALT) or aspartate aminotransferase (AST) as a proxy for hepatic steatosis) were excluded. Hypocaloric diets are defined as containing less calories than required for energy demands and usually contain 1000–1200 kcal/day. Very low-calorie diets (VLCDs) typically contain less than 800 kcal/day. See the appendix for literature search details.

ADIPOSE TISSUE DYSFUNCTION

The cause for adipose tissue dysfunction and ectopic fat storage is largely unknown. Blüher [10] recently proposed a model in which genetic, environmental, and behavioral factors are involved in excess energy intake and decreased physical activity on one hand as well as physiologic versus pathologic fat accumulation on the other hand. The physiologic response to excess energy intake and less expenditure would be an increase in fat cell number as well as fat cell size (hyperplasia of adipose tissue) in the subcutaneous compartment. These fat cells would function normally and be insulin sensitive. On the contrary, a pathophysiologic response would be adipocyte hypertrophy. Human adipocytes can grow up to ~20 fold in diameter and several thousand-fold in volume [14]. However, when adipocytes become too large, stress signals will be released. For example, hypoxia can occur when vascularisation is inadequate for the expanded adipose tissue. In addition, endoplasmatic reticulum (ER) stress [15], either induced by hypoxia or nutrient excess, leads to an unfolded protein response (UPR) [16]. In the ER, proteins are translated, folded, and assessed for quality before release. With ER stress the number of misfolded proteins increases and triggers the UPR. The UPR induces genes involved in assembling, folding, modifying, and degrading proteins to alleviate ER stress and triggers the activation of stress and inflammatory pathways and the production of cytokines and chemokines that interfere with the insulin signaling pathway.

Dysfunctional adipocytes produce more inflammatory adipokines and cytokines than functional adipocytes. In addition, stressed adipocytes attract immune cells (among which macrophages) into the stromal vascular fraction. The physiological role of infiltrating macrophages is probably debris clearing (ER stress can eventually lead to premature adipocyte apoptosis). Eventually however, a positive feedback cycle is formed in which infiltrating macrophages recruit more immune cells, and a state of chronic inflammation is induced. Some of the cytokines and adipokines produced interfere with adipocyte differentiation, others with insulin signaling. Some, like TNF-α and IL-6, impair adipocyte differentiation, reduce lipid accumulation, and increase adipocyte lipolysis.

The latter is on top of the poor ability of stressed and hypertrophic adipocytes to take up and release free fatty acids (FFAs). This induces a redirection of lipids towards peripheral tissues, including skeletal muscle, liver, pancreas, and heart: ectopic fat deposition.

If in these tissues, lipid supply exceeds oxidative capacity intracellular lipid accumulation occurs, and organ function might be impaired (Figure 1).

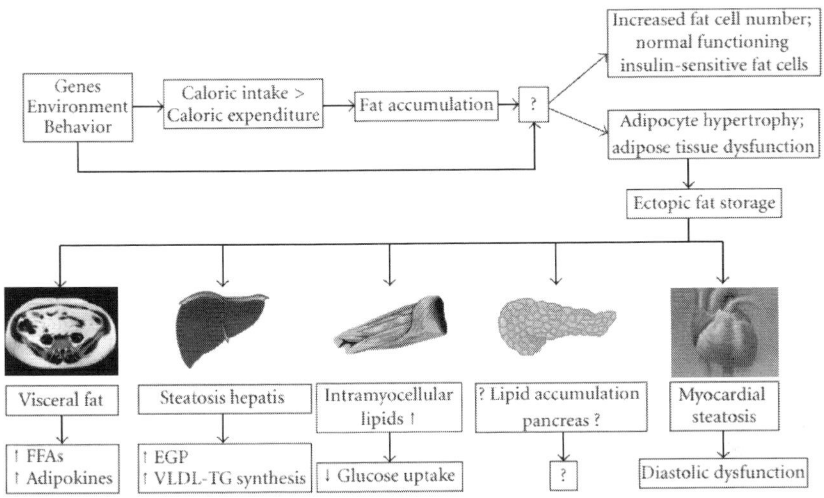

Figure 1: Ectopic fat depositions. Gene-environment interactions might be involved both in increased energy intake and decreased physical activity as well as in the response of adipose tissue to the ensuing increased energy balanced. Predisposed subjects will elicit a pathophysiologic response leading to adipocyte hypertrophy. This will lead to an inflammatory response that ultimately leads to ectopic fat deposition. The consequences of ectopic fat are organ specific as depicted in this figure. The effect of ectopic fat on the cellular level is different and depicted in Figure 2. FFAs: free fatty acids; EGP: endogenous glucose production; VLDL-TG: very low-density lipoprotein-triglyceride.

CELLULAR MECHANISMS OF ECTOPIC FAT-INDUCED ORGAN DYSFUNCTION

The consequences of ectopic fat accumulation depend on the specific organ. However, the mechanisms leading to the disruption of organ function are quite similar at the cellular level (Figures 2(a)–2(c)).

Firstly, it has to be noted that lipids can be dispersed intercellularly or accumulate intracellularly. Intercellular lipid accumulation might impair organ function via paracrine effects of the released adipokines. However, it is intracellular lipid accumulation that is associated with decreased insulin sensitivity.

Free fatty acids are taken up by the cell via specific transport proteins in the cell membrane (CD36, fatty acid transport protein (FATP), along with passive diffusional uptake [17]. Inside the cell, fatty acid binding protein

(FABPc) is the most important cytosolic protein for guiding long-chain fatty acids in the cell to places of oxidation or esterification. Long-chain fatty acyl-CoA is taken up by the mitochondria via carnitine-palmitoyl transferase 1 (CPT1). Inside the mitochondria, β-oxidation and further degradation in the tricarboxylic acid cycle takes place. Therefore, intracellular lipid accumulation occurs as a consequence of continuous oversupply of FFAs (caused by enhanced lipolysis, adipocyte dysfunction) together with an impairment in FFA oxidation in the mitochondria.

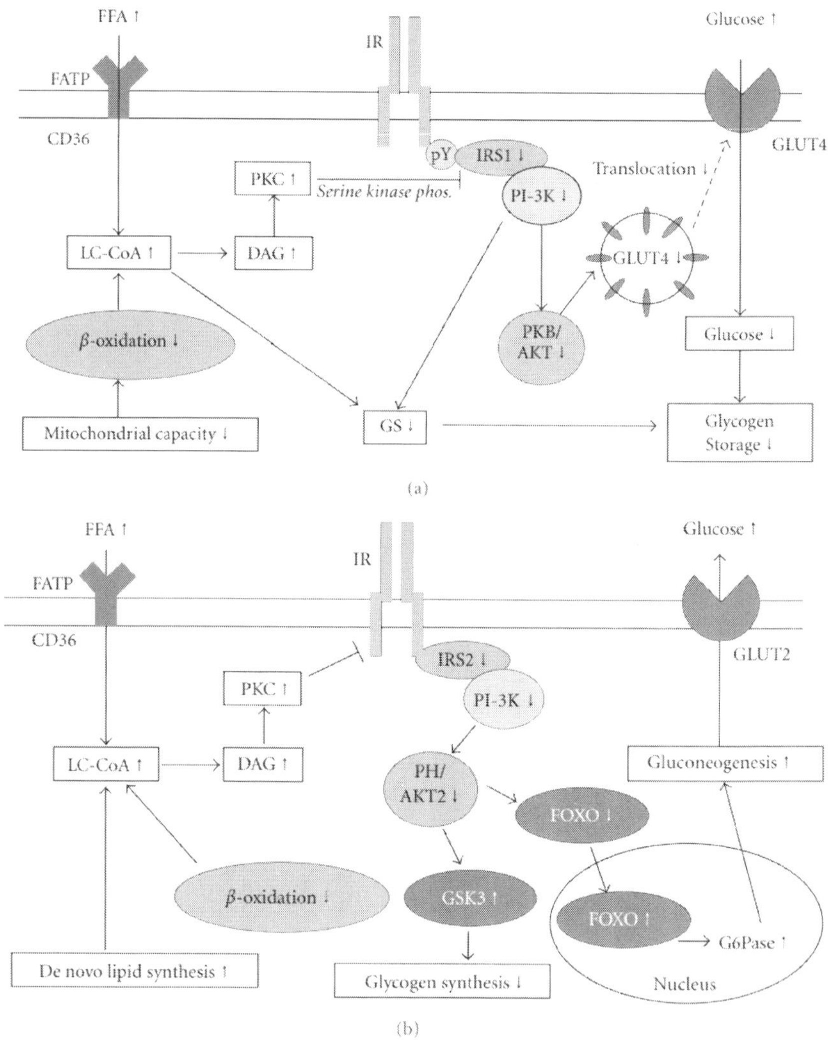

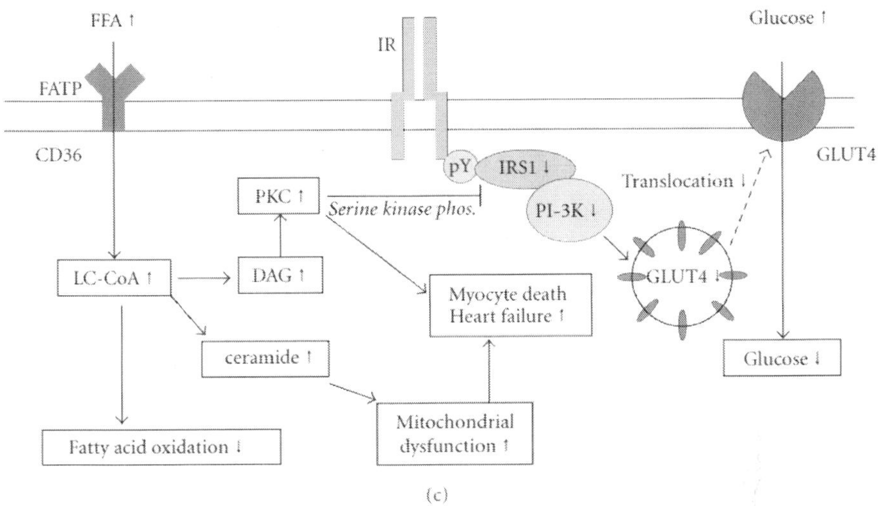

Figure 2: (a) cellular processes leading to insulin resistance in the skeletal muscle cell; (b) cellular processes leading to insulin resistance in the liver cell; (c) cellular processes in the cardiomyocyte leading to myocardial dysfunction.

It appears that not the FFAs themselves but rather metabolites like long-chain acyl-CoA (LC-CoA), diacylglycerol (DAG), and ceramides are deleterious for the cell. These fatty acid metabolites induce a sustained activation of serine/threonine kinases such as protein kinase C (PKC) isoforms, IKB-kinase-β and Jun N-terminal kinase, which phosphorylate insulin-receptor substrates (IRS) on serine residues [18]. The subsequent defects in insulin signaling lead to a decrease in cellular function that depends on the cell type. Figure 2describes the effects of intracellular lipid accumulation in skeletal muscle, liver, and cardiomyocytes [18, 19]. The explanation for the various organs is described in the following paragraphs and is the explanation of Figures 2(a) to 2(c).

FFAs are taken up by the skeletal muscle cell mainly by protein-mediated membrane transport (CD36, fatty acid transport protein (FATP)), along with passive diffusional uptake. Long-chain fatty acyl-CoA is taken up by the mitochondria via carnitine-palmitoyl transferase 1 (CPT1) [17]. Inside the mitochondria β-oxidation and further degradation in the tricarboxylic acid cycle takes place. When there is an oversupply of FFAs and/or an impairment in mitochondrial β-oxidation exist, LC-CoA accumulates and is broken down to intermediates like DAG and ceramide. These fatty acid metabolites induce a sustained activation of serine/threonine kinases such as protein kinase C (PKC) isoforms, IKB-kinase-β and Jun N-terminal kinase, which phosphorylate insulin-receptor substrate (IRS) 1 on serine residues [18]. Serine-phosphorylated forms of IRS1 cannot associate with and activate phosphatidyl-inositol-3-kinase (PI3K), resulting in a decreased glucose transporter 4 (GLUT4) regulated glucose transport over the cell membrane. FATP: fatty acid transport protein; FFA: free fatty acid; LC-CoA: long-chain fatty acid-CoA; DAG: diacylglycerol; PKC: protein kinase C; IR: insulin receptor; IRS: insulin receptor substrate; PI3K: phosphatidylinositol 3-kinase; PKB/AKT: protein kinase B/AKT; GLUT4: glucose transporter protein 4; GS: glycogen synthase.

The mechanism behind hepatic TG accumulation and the development of hepatic insulin resistance are similar to that described for skeletal muscle [18, 20]. An increase in DAG in hepatic cells leads to activation of PKC, leading to decreased insulin receptor kinase activity and subsequently lower insulin-stimulated IRS2 tyrosine phosphorylation and lower IRS2-associated PI3K-activity which results in reduced insulin stimulation of glycogen synthase activity. This ultimately leads to decreased insulin-stimulated hepatic glucose uptake and reduced insulin suppressibility of hepatic glucose production. Furthermore, reduced activity of AKT2, a protein kinase downstream of IRS and PI3K, results in decreased phosphorylation of the forkhead box O (FOXO) transcription factor, allowing it to enter the nucleus and activate the transcription of the rate-controlling enzymes of gluconeogenesis. The result is increased hepatic glucose production and decreased hepatic glucose uptake, which both contribute to increased plasma glucose levels. FATP: fatty acid transport protein; FFA: free fatty acid; LC-CoA: long-chain fatty acid-CoA; DAG: diacylglycerol; PKC: protein kinase C; IR: insulin receptor; IRS: insulin receptor substrate; PI3K: phosphatidylinositol 3-kinase; GSK3: glycogen synthase kinase 3; FOXO: forkhead box protein O; GLUT2: glucose transporter protein 2.

In the heart, increased FFA supply and decreased oxidation lead to lipid intermediates that impair insulin signaling, and hence GLUT-4 involved glucose uptake. In addition, lipid-induced PKC activation might be involved in premature adipocyte apoptosis and heart failure. Finally, accumulation of lipids increase DAG and ceramide levels inducing a vicious cycle in which a further impairment in mitochondrial fatty acid oxidation occurs that also has negative effects on myocardial function. FATP: fatty acid transport protein; FFA: free fatty acid; LC-CoA: long-chain fatty acid-CoA; DAG: diacylglycerol; PKC: protein kinase C; IR: insulin receptor; IRS: insulin receptor substrate; PI3K: phosphatidylinositol 3-kinase; GLUT4: glucose transporter protein 4.

SKELETAL MUSCLE

Intramyocellular Lipids and Peripheral Insulin Resistance

Accumulation of intramyocellular lipids (IMCLs) as measured by skeletal muscle biopsies or ^{1}H-magnetic resonance spectroscopy is associated with insulin resistance [21–26] and T2DM. However, it is not synonymous with the condition given the fact that endurance-trained athletes, who are highly insulin sensitive, also have a high IMCLs content [26]. Rather, the capacity to oxidize IMCLs determines whether they represent a physiological or a pathological role [27]. In endurance-trained athletes IMCLs are an adaptive response, the IMCLs serve as a readily available energy source. The close proximity of lipid droplets to the mitochondria supports this hypothesis. In these athletes, IMCLs are not deleterious because of the increased capacity to oxidize lipids. In insulin-resistant and/or T2DM patients, the increased IMCLs stores are the result of increased FFA availability and impaired fatty acid oxidation [28–30]. The latter leads to accumulation of lipid intermediates with the ascribed toxic effects on insulin signaling (Figure 2). Hence, the ratio between IMCLs and fat oxidative capacity is a better marker for insulin resistance than IMCLs alone.

The fact that fatty acid oxidation is impaired in insulin-resistant states led to the speculation that mitochondrial dysfunction is the cause of IMCLs accumulation and the ensuing insulin resistance. Indeed, a decreased mitochondrial density and/or function has been reported in insulin-resistant offspring of T2DM patients [31–33] and T2DM patients [34–37]. However, three of these studies reported decreased mitochondrial function at normal IMCLs levels [35–37] suggesting that impaired mitochondrial function is not a prerequisite for IMCLs accumulation. Rather, it might be that mitochondrial dysfunction is the consequence of the increased amount of fatty acid metabolites, for example, via the formation of lipid peroxides [38]. In that case it might be that the lipid-induced mitochondrial dysfunction induces progressive deterioration of oxidative capacity and further accumulation of lipid intermediates in the skeletal muscle cell. Further investigations are warranted to elucidate which one is cause or consequence: IMCLs or mitochondrial dysfunction. In this paper, we focus on IMCLs and will not elaborate on mitochondrial function.

Effect of Diets on IMCLs Accumulation (Table 1)

Table 1: Effect of diet and exercise on insulin sensitivity and intramyocellular lipid (IMCLs) content

Ref.	Patients	n	BMI start kg/m²	Age yrs	Intervention	Duration	Body weight loss kg	Effect on skeletal muscle insulin sensitivity	Effect on IMCLs
[45]	Obese NGT	12	31 ± 2	Unknown	25% caloric restriction	6 months	−8 kg	Si: no change	No change
		12	33 ± 2		12.5% caloric restriction + 12.5% exercise		−8 kg	Si 37 ± 18%, $P < 0.01$	No change
		11	33 ± 2		15% weight loss hypocaloric diet 1200 kcal/day		11 kg	Si 70 ± 34%, $P < 0.04$	No change
		11	31 ± 2		controls		0 kg	Si: no change	No change
[39]	Morbid obese NGT	9	48 ± 9	39 ± 12	Diet 1200 kcal/day	6 months	−14 ± 12 kg	M value: no change	No change
		8	51 ± 8	39 ± 12	Biliopancreatic diversion		−33 ± 10 kg	M value 23 ± 3 to 52 ± 11 μmol/kg FFM/min, $P < 0.05$	1.6 ± 1.1 to 0.2 ± 0.4 AU, $P < 0.05$
	Controls	7	27 ± 1	35 ± 11	Controls		Unknown	M value: no change; baseline 53 ± 13 μmol/kg FFM/min	No change; baseline 0.1 ± 0.2
[46]	Obese NGT	20	34 ± 1	42 ± 2	700 kcal compared to normal diet	15 weeks	11 kg	OGTT: no change	No change
					followed bij energy restriction + exercise	21 ± 2 weeks	−5 kg	OGTT: no change	No change
[41]	Obese T2DM	8	30 ± 1	47 ± 3	1200 kcal/day 3% fat diet (untill normoglycemia)	3–12 weeks (mean 7)		GDR: no change	No change
					studies after weight stabilisation period	4 weeks	−8 ± 1 kg		
[42]	Obese NGT	13	33 ± 2	Unknown	−522 kcal compared to normal diet	3 months	−6 kg	M value: no change	No change
[43]	Overweight T2DM	7	27 ± 3	55 ± 5	−25–30 kcal/kg LBM	2 weeks	BMI −1.5 ± 0.0%	M value: no change	No change
		7	27 ± 3	46 ± 3	−25–30 kcal/kg LBM + advice to walk 2-3 ld 5-6 days/week		BMI 2.3 ± 0.1%	M value 5.5 ± 0.3 to 8.2 ± 0.5 mg/kg/min; $P < 0.001$	3.8 ± 0.4 to 3.1 ± 0.4 IMCL/Cr, $P < 0.03$
[44]	Morbid obese NGT	7	44 ± 6	Unknown	Diet 1200 kcal/day	6 months	−5 ± 4 kg	M value: no change	No change
[48]	Obese T2DM	13	36 ± 1	50 ± 3	VLCD 600–800 kcal/day	8 weeks	−9 ± 1 kg	HOMA-IR −0.9 is −44 ± 7%, $P < 0.001$	No change
[49]	Obese NGT	5	36 ± 5	38 ± 12	VLCD 700 kcal/day	6 days	2.3 kg	GDR: no change	−56%, $P = 0.006$ (^{1}H-MRS)
	Obese T2DM	7	37 ± 7	43 ± 6	VLCD 700 kcal/day	6 days	−3.7 kg	GDR: no change	−40%, $P = 0.04$ (^{1}H-MRS)
[50]	Obese T2DM	10	40 ± 2	55 ± 3	VLCD 500 kcal/day until 50% excess weight was lost	Mean 17 weeks	−22 kg	GDR 18.8 ± 2.0 to 39.1 ± 2.8 umol/kg LBM/min, $P < 0.001$	7 ± 14 to 4 ± 1 AU, $P < 0.002$
[51]	Obese NGT	7	33 ± 1	46 ± 2	25% caloric restriction, goal 7% weight loss	18.6 ± 0.7 weeks	−11 ± 2%, ca −8 kg	M value increased 29 ± 7% $P < 0.05$	Decreased, $P < 0.05$
		9	35 ± 1	42 ± 3	25% caloric restriction +3/5 days/wk exercise at 60–70% MHR	19.2 ± 0.4 weeks	9 ± 1%, ca −9 kg	M value increased 38 ± 9% $P < 0.05$	No change
[52]	Obese T2DM	10	34 ± 1	44 ± 3	25% caloric restriction +3/5 days/wk exercise at 60–70% MHR	16–20 weeks	7.1 ± 0.1% = ca 7 kg	GDR 4.1 ± 0.6 to 6.3 ± 0.9 mg/kg LBM/min, $P < 0.05$	48 ± 1 to 50 ± 1 HU, $P < 0.01$ (CT)
[53]	Obese NGT	21	33	40	−500–1000 kcal compared to normal diet with 4–6 x/wk exercise at 65–75% MHR	16 weeks	−10 kg	GDR 6.5 naar 9.7 mg/kg LBM/min $P < 0.05$	No change
[54]	Obese IGT	11	34 ± 1	67 ± 1	−500 kcal/day compared to normal diet and aerobic exercise 5 days/wk 60 min at 75% VO_{2max}	12 weeks	ca −8 kg	M value 2.9 ± 0.3 to 4.7 ± 0.6 mg/kg LBM/min, $P < 0.01$	3.9 ± 0.6 to 2.5 ± 0.3 LAI%, $P < 0.05$
		12	35 ± 2	66 ± 1	Aerobic exercise 5 days/wk 60 min at 75% VO_{2max}	12 weeks	ca −3 kg	M value 3.0 ± 0.4 to 4.2 ± 0.7 mg/kg LBM/min, $P < 0.05$	3.9 ± 0.6 to 3.0 ± 0.4 LAI%, $P < 0.05$
[55]	Obese NGT	25	30 ± 1	66 ± 1	4-5 days/wk supervised aerobic exercise at 75% MHR	16 weeks	−1.3 kg	M value: no change	21% increase, $P < 0.01$
[56]	Obese NGT	20	30 ± 1	59 ± 1	2x/week 30 min aerobic + 1x/week resistance exercise both at 55% VO_{2max}	12 weeks	no change	GDR: no change	No change
	Obese T2DM	18	30 ± 1	59 ± 1			no change	GDR 18.4 ± 1.4 to 21.0 ± 1.4 umol/kg/min, $P < 0.05$	No change
[47]	Obese IGT	8	31.2 ± 1.2	66.6 ± 1.2	Diet aimed at 10% weight loss and <30% fat	16 weeks	−8.2 ± 1.2%	Insulin-stimulated Rd + 20.6 ± 4.8%, from ~8 mg/kg FFM/min	Decrease 16.0 ± 3.2%
	or IGT	8	30.0 ± 1.0	68.4 ± 1.2	Ex: 4-5 days/week 45 min/session moderate intense	16 weeks	−1.8 ± 0.6%	Insulin-stimulated Rd + 16.2 ± 12.6%, from ~8 mg/kg FFM/min	Increase 40.8 ± 18.2%

NGT: normal glucose tolerant; GDR: glucose disposal rate; HOMA-IR: homeostatic model assessment of insulin resistance; IGT: impaired glucose tolerant; LBM: lean body mass; T2DM: type 2 diabetes mellitus; AU: arbitrary units; LAI: mean percentage area of an individual skeletal muscle fibre that stains for lipids.

When assessing the effects of diet on insulin sensitivity and IMCLs content both severity of the caloric restriction, duration of the intervention, amount of weight loss and perhaps severity of insulin resistance and duration of T2DM have to be taken into account. From the available literature (Table 1), it is difficult to draw definite conclusions as to which of these parameters has the greatest effect on improving insulin sensitivity and IMCLs [39–45].

For example, moderate weight reduction (3–11 kg) with hypocaloric diets showed no effect on insulin sensitivity or on IMCLs content in normal glucose tolerant (NGT) obese subjects [39, 42, 44–46] or T2DM patients [41, 43] in 3 studies. However, a recent study did show an increase in insulin-stimulated glucose disposal along with a decrease in IMCLs content following around 8 kg weight reduction with a hypocaloric diet (−500 to –40 kcal/day based on food diaries) in elderly obese subjects with impaired fasting glucose or impaired glucose tolerance [47]. Maybe the low fat content of the diet accounts for the observed difference as compared to the other 3 studies. Alternatively, the pathogenesis of impaired glucose metabolism (and hence its reversal) might be different in older as compared to younger subjects.

The amount of caloric restriction seems to be important since moderate weight reduction with a low-calorie diet [45] or a very low-calorie diet [48] significantly improved insulin sensitivity although it had no effect on IMCLs in obese NGT subject. One study using a VLCD showed not only an improvement in insulin sensitivity but also a decrease in IMCLs, even after 6 days of the VLCD, in obese and T2DM patients [49]. It should be noted; however, that IMCLs were measured in soleus muscle with ^{1}H-MRS, whereas other studies mostly measure IMCLs in the vastus lateralis muscle. In addition, with respect to insulin sensitivity, very high doses of insulin (200 mU/m^{2}/min) were used.

Pronounced weight loss (~22 kg) using a VLCD (450 kcal/day, on average 17 weeks duration) in obese insulin-treated T2DM patients not only improves peripheral insulin sensitivity but also decreases IMCLs in skeletal muscle biopsies [50] (Table 1).

Effect of Diet and Exercise on IMCLs Accumulation

Several studies investigated the effect of the addition of exercise to a hypocaloric or VLCD in obese NGT or obese T2DM patients (Table 1). All studies in obese NGT [45, 51, 53], obese impaired glucose tolerant (IGT) [54], or obese T2DM patients [43, 52] consistently showed increased insulin sensitivity with no differences between diet-only or diet with exercise group (with the exception of [45] all measured with the hyperinsulinaemic euglycaemic clamps). The effect of adding exercise on IMCLs in these studies was less consistent showing either no effect [45, 51, 53], a decrease [43, 54], or even an increase [52] in IMCLs accumulation.

Effect of Exercise on IMCLs Accumulation

Few studies have addressed the effect of exercise only on IMCLs and insulin sensitivity. Three studies on obese NGT [55] and obese IGT subjects [47, 54] showed an increase in insulin-stimulated glucose disposal despite minimal weight loss (1–3 kg). Two of those studies [47, 55] showed an increase in IMCLs with a decrease in intramuscular ceramide and DAG levels along with increased lipid oxidation rates. The other study showed a decline in IMCLs both in the diet + exercise as well as in the exercise-only group [54].

In 18 T2DM patients (body mass index (BMI) 30 kg/m^2, HbA1c 7.2%, age 59 years) a 12-week combined aerobic and resistance program had no effect on body weight [56] but improved insulin-stimulated glucose disposal and increased IMCLs along with an increase in insulin-stimulated glucose oxidation and suppression of fat oxidation (Table 1). Interestingly, the same intervention had no effect on peripheral insulin sensitivity and IMCLs in healthy-matched controls.

CONCLUSION

Moderate weight loss has no effect on IMCLs but can improve peripheral insulin sensitivity if a VLCD is used. Larger weight losses are needed to improve both peripheral insulin sensitivity and deplete IMCLs stores. With respect to the effect of exercise on IMCLs, the story is more complex. Athletes have increased IMCLs stores that serve as a readily available energy supply and training further increases IMCLs in healthy subjects. On the other hand, sedentary insulin-resistant subjects also have increased IMCLs, caused by diminished oxidative capacity. These are associated with increased intramyocellular lipid intermediates that interfere with insulin signaling and other cellular processes. Exercise training in these subjects either increases, decreases, or has no effect on IMCLs accumulation. This is probably associated with the stage of disease (can fatty acid oxidation and/or mitochondrial function be restored) as well as with the intensity of the exercise program and sort (aerobic/resistance) exercise. Overall, studies show that exercise improves peripheral insulin sensitivity. Factors involved are a reduction in lipid intermediates and hence improved insulin signaling and mitochondrial function, increased oxidative capacity and increased capillary blood flow. In addition, contraction and hypoxia activate adenosine monophosphate-activated protein kinase (AMPK) which leads to increased GLUT4 translocation independent of insulin.

LIVER

Intrahepatic Lipids and Insulin Resistance of the Liver

Accumulation of fat in the liver in the absence of excessive alcohol ingestion is referred to as nonalcoholic fatty liver disease (NAFLD). The spectrum of liver abnormalities within this entity ranges from hepatic steatosis with or without mild increases in serum AST/ALT to nonalcoholic steatohepatitis (NASH) with or without fibrosis, cirrhosis, and incidental hepatocellular carcinoma. The worldwide estimated prevalence of NAFLD in the general population is about 20%, with large differences across countries [57]. There is a strong positive correlation between NAFLD and obesity. A cross-sectional study on 1515 severely obese NGT subjects showed some signs of fatty liver changes in almost 90% of people [58]. The majority had hepatic steatosis, but one-third had portal inflammation and fibrosis. A prospective study demonstrated a 4-time increased risk to develop hepatic steatosis in obese persons as compared to controls [59].

Noninvasive methods for measuring hepatic TG content such as ultrasound, CT, and ^{1}H-MRS cannot distinguish NAFLD from NASH and fibrosis. A definite diagnosis can only be made by liver biopsy with histologic examination. The cutoff value for abnormal lipid accumulation in the liver has been defined as more than 5% of liver volume or when more than 5% of hepatocytes contain visible intracellular lipids [60]. Two recent studies in respectively a NGT mixed (Hispanic, non-Hispanic, and African American) population [61] and a lean Caucasian population [62] found that the 95th percentile for hepatic TG content was 5.6% and 3%, respectively, using ^{1}H-MRS. The Pathology Committee of the NASH Clinical Research Network designed and validated a scoring system of 14 histological features examining liver biopsy findings detailing steatosis, fibrosis, inflammation, and liver cell injury [63]. An NAFLD activity score >5 was universally associated with NASH.

Fat accumulation in the liver is associated with hepatic insulin resistance as well as with peripheral insulin resistance in skeletal muscle and adipose tissue [64–67]. In a large European cohort of 1307 middle-aged NGT subjects, patients with a high fatty liver index (an estimate for hepatic steatosis based on an algorithm including BMI, waist circumference, TG, and gamma-glutamyltransferase) had a lower glucose disposal rate as measured by a euglycaemic hyperinsulinaemic clamp as well as higher FFA levels at the end of the insulin infusion [68]. The latter is suggestive for decreased insulin sensitivity of adipose tissue. Korenblat et al. [64] found a negative correlation between hepatic insulin sensitivity measured by the hyperinsulinaemic euglycaemic clamp and hepatic TG content (measured by ^{1}H-MRS) in 42 nondiabetic obese subjects. Indeed, a multivariate linear regression analysis found that hepatic TG content was the best predictor of insulin sensitivity in liver, skeletal muscle, and adipose tissue, independent of BMI and percent body fat. Some claim that the amount of hepatic TG content is directly correlated with the severity of insulin resistance, but this cannot be confirmed by others.

Effect of Diet on Intrahepatic Lipids

Noninvasive techniques like CT and ^{1}H-MRS have shown that weight loss by nutritional interventions can result in a large decrease in hepatic TG content in obese and T2DM subjects [41–43, 69–73]. Because of the different populations and differences in baseline hepatic TG content, studies are not well comparable with respect to the individual effect of level of caloric restriction and/or amount of weight reduction on loss of hepatic TG content. Nonetheless, it has been shown that even a relatively small drop in BMI of 3–6% is associated with a considerable reduction in hepatic TG content of 34–40% [42, 70, 71]. The main reduction in hepatic TG content already occurs in the first two weeks of dietary restriction [43, 69]. The percentual decline in hepatic TG content positively correlates with the hepatic TG content at baseline, as patients with a high hepatic TG content at start of the diet lose relatively more TG than patients with low hepatic TG content, with the same amount of weight loss [70, 72]. Only two studies measured both hepatic TG content and hepatic insulin sensitivity with a hyperinsulinaemic euglycaemic clamp in obese NGT [73] and obese T2DM patients, respectively [41]. Both studies found an improvement in hepatic insulin sensitivity that was associated with the decrease in intrahepatic lipids (IHL).

Although the above-mentioned studies clearly show that diet-induced weight loss leads to a decrease in hepatic TG content, the effect of this decrease in hepatic TG content on liver histology (i.e., with liver biopsies) has only been scarcely studied. In obese patients with NASH moderate weight loss over a 12-month period, obtained with dietary advices only, led to an improved steatosis score in 9 of the 15 subjects. The improvement was associated with greater weight loss as compared to the patients that showed no change in steatosis [74]. Another study on 15 patients with obesity and NASH combined a hypocaloric diet with exercise during 12 weeks, while 10 obese subjects with NASH served as controls. The BMI decreased by 3 kg/m^2 in the intervention group, and the steatosis score decreased from 2.3 to 1.3 (30–50% steatosis to less than 30%), while no changes were observed in the control group [75]. In contrast, severe caloric restriction during 8 months in 41 morbidly obese (BMI 43.3 kg/m^2) subjects leading to an impressive median weight loss of 34 kg (−27% of BMI) showed normalization of liver architecture in 19 patients [76]. However, they also found an increase in hepatic inflammation and fibrosis in some patients, this was associated with a greater weight loss and elevated FFAs.

Effect of Diet and Exercise on Intrahepatic Lipids

Most studies combine an exercise program with a hypocaloric diet. Since short-term diet alone already reduces IHL, it is difficult to extrapolate whether adding exercise has additional effects [77, 78]. Both in obese as well as in T2DM patients combined diet with exercise interventions led to a reduction of hepatic steatosis. Larson-Meyer studied a diet-only (25% calorie restriction (CR)) versus a diet (12.5% CR) combined with exercise (12.5% CR) obese NGT group. In this 6-month study no additional effect of exercise upon the diet was found at an equal total amount of caloric restriction [45]. Another 6-month study in older sedentary obese subjects also showed no additional effect of exercise on IHL content and insulin sensitivity measured by oral glucose tolerance test (IHL diet: −46 ± 11%, diet + exercise: −45 ± 8%; insulin sensitivity diet: 66 ± 25%, diet + exercise: 68 ± 28%) [79].

Effect of Exercise on Intrahepatic Lipids

Only one study has investigated the influence of exercise alone on hepatic TG content in adult overweight (BMI 27.7 ± 0.5 kg/m^2) sedentary men [80] and simultaneously measured hepatic insulin sensitivity with a state of the art technique. After a 6-week aerobic exercise program (60–85% of $VO_{2\,max}$ for a minimum of 20 min. at least three times per week) without significant effect on body weight no changes were found in hepatic TG content as measured by ^{1}H-MRS, although both peripheral and hepatic insulin sensitivity (measured by the hyperinsulinaemic euglycaemic clamp) improved.

This is in contrast with a study in obese NGT subjects in which a one-month aerobic exercise training (3 times/week, 3–45/min at max 70%) decreased hepatic TG concentration by 21% at equal body weight. Hepatic insulin sensitivity was not measured, HOMA-IR remained unchanged [81].

Perhaps the amount of IHL at start of the study and/or the severity of the impairment in glucose metabolism accounts for the observed differences.

Interestingly, in 196 overweight, sedentary dyslipidemic subjects an 8-month exercise intervention showed that aerobic exercise is more effective in reducing liver fat and insulin sensitivity (measured by HOMA) than resistance training. A combined resistance and aerobic exercise program was as effective as aerobic exercise on the above-mentioned outcome parameters [82].

EPICARDIAL AND PERICARDIAL FAT

Pericardial fat is the adipose tissue surrounding the heart. It consists of two layers: epicardial fat (visceral fat) originating from mesothelial cells and paracardial or mediastinal fat, originating from mesenchymal cells [83] (Figure 3).

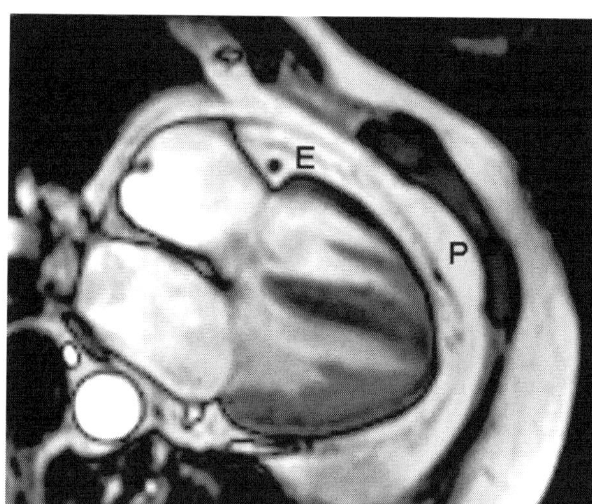

Figure 3: Four-chamber view of the heart. Four-chamber view of the heart where the signal from blood and muscle are suppressed. E: epicardial fat; P: pericardial fat.

Around 80% of the heart is surrounded by epicardial fat and epicardial fat makes out 20% of total heart weight. The latter varies more among men than in women; 15.2–25.2% and 19.5–21.7%, respectively. The largest part can be found around the right ventricle followed by the anterior wall. The amount of epicardial fat increases with age until age 20–40, thereafter there is no age dependence [83, 84].

Several functions have been proposed for epicardial fat tissue [84]. Scientific proof is, however, rather difficult to obtain since most animal species have very little epicardial fat [85]. In guinea pigs, rates of lipolysis and lipogenesis were 2 fold higher in epicardial fat than in other fat depots. This led to the assumption that epicardial fat might act as a buffer to protect the myocardium from highly toxic fatty acid levels and to provide fatty acids as a direct energy source in times of energy demand [85, 86]. Coronary arteries are embedded in epicardial fat so that another putative function might be to protect the coronary arteries from the tension and torsion induced by the arterial pulse wave and provide an environment in which the coronary arteries can easily expand. This fat compartment also acts as a metabolically active organ, secreting cytokines [87, 88].

Several cross-sectional studies have suggested a positive relation between an increased epicardial fat volume and coronary artery disease [89–92]. Furthermore, an increased epicardial fat volume has been associated with insulin resistance in nondiabetic obese patients [93] and with the presence of T2DM in a Han Chinese population [94, 95]. Therefore, it is interesting to study the effects of weight loss and diet on epicardial fat and insulin resistance.

Effect of Diet on Epicardial Fat

Two studies have examined the effect of diet-induced weight loss on epicardial fat. Kim et al. [96] studied 27 moderately obese NGT subjects, who lost 11% (9.5 kg) of weight during a 12-week weight loss intervention study. Epicardial fat thickness measured over the right ventricle wall by echocardiography decreased by 17% from baseline. Iacobellis et al. [97] studied 20 severely obese (BMI 45 ± 5 kg/m^2) subjects (probably including patients with IGT or T2DM) who followed a 6-month low-calorie diet (900 kcal/day) and lost 20% (25 ± 10 kg) of bodyweight. Epicardial fat decreased by 32% from baseline. This was accompanied by an improvement in left ventricular mass and diastolic cardiac function. The change in diastolic function was also positively correlated with the change in epicardial fat thickness. Moreover, in 15 obese patients with T2DM, we found a significant decrease in pericardial fat after weight loss with a 16-week VLCD, measured with MRI (accepted for publication in Obesity) [98].

Effect of Exercise on Epicardial Fat

To date, only two studies examined the effect of (aerobic) exercise on epicardial fat [99, 100]. In one study, 24 obese NGT (BMI 30.7 ± 3.3 kg/m^2) middle-aged Japanese men followed a supervised exercise program for 3 months. The exercise intensity was gradually increased in 4 weeks from 50–60 to 60–70% of the maximum heart rate 3 days/week 60 minutes, which was continued for the remainder of the study. Following the intervention the BMI decreased by $4.3 \pm 3.0\%$ (circa -1 kg/m^2) and $VO_{2\,max}$ increased by 20%. Epicardial fat thickness measured by echocardiography over the free wall of the right ventricle decreased significantly. The change in visceral adipose tissue (−15%) was significantly correlated with the change in epicardial adipose tissue (−8.6%).

In the other study, 32 obese postmenopausal women were randomized to diet-only or diet combined with moderate or intensive exercise for 20 weeks. All three groups had a similar 15% reduction in bodyweight and a 17% reduction in pericardial fat. However, no differences were observed between the diet-only and diet with exercise group [100].

MYOCARDIAL TRIGLYCERIDE CONTENT

In addition to the epicardial/pericardial fat depositions, TG can also be stored within the cardiomyocytes. This is referred to as myocardial TG content or myocardial steatosis. Myocardial TG accumulation can be measured with great sensitivity by ^{1}H-MRS [101]. Patients with IGT and T2DM have an increased myocardial TG content compared to obese and lean controls [102–104]. This accumulation of myocardial TG is a result of excessive fatty acid uptake relative to the oxidation. If fatty acids are converted to myocardial TG, several intermediates are released (e.g., ceramide) which in animal models caused cardiac dysfunction [105, 106]. In T2DM patients, myocardial steatosis was associated with impaired left ventricular diastolic function [104]. Indeed, an increased fatty acid uptake in the myocardium has been found in healthy obese and T2DM patients compared to healthy lean subjects [107, 108]. But in the one study that investigated the fate of the intracellular fatty acids in the resting state in insulin-naïve T2DM patients as compared to controls an increase, not a decrease in fatty acid oxidation was found. Fatty acid reesterification was negligible but lower in T2DM patients as compared to the controls [108]. Interestingly, neither plasma NEFA levels nor myocardial blood flow was increased both in subjects with obesity [107] and T2DM [108] suggesting another mechanism, for example, at the (cellular) level of the FAT/CD36, to account for the increased fatty acid uptake (Figure 2(c)).

The myocardial TG content is not static. Three days of severe caloric restriction (450 kcal/day to complete starvation) in healthy volunteers and patients with T2DM increases myocardial TG content, which is associated with a decrease in left ventricular diastolic function [109, 110].

Effect of Diet on Myocardial Triglyceride Content

Thirtyfour obese (BMI 33.7 ± 0.7 kg/m^2) healthy persons underwent a 6-week VLCD (550 kcal/day). Before and after the intervention intramyocardial TG were measured, a hyperinsulinaemic euglycaemic clamp was performed and either glucose uptake or fatty acid uptake was measured by positron emission tomography. The intervention led to a weight loss of 11.2 ± 0.6 kg and a nonsignificant decrease in myocardial TG content of 31% ($n = 8$, $P = 0.076$). Myocardial fatty acid uptake decreased significantly. Myocardial mass and work decreased significantly by 7 and 26%, respectively [111].

Our group investigated the effect of a 16-week VLCD (450 kcal/day) on myocardial TG content in obese patients with T2DM [112]. We showed that a decrease in BMI (from 35.6 ± 1.2 to 27.5 ± 1.3 kg/m^2) was associated with a significant decrease in myocardial TG content and an improvement in left ventricular diastolic function.

Effect of Exercise on Myocardial Triglyceride Content

In 14 NGT overweight-obese middle-aged men, a 12-week training program (3 times per week, combined aerobic and resistance training) reduced lipid accumulation at the cardiac septum (0.99 ± 0.15 to 0.54 ± 0.04%, $P = 0.02$) and improved left ventricular function slightly but significantly [113].

In obese patients with T2DM, however, a comparable 12-week intervention had no effect on cardiac lipid content measured in the septum, although both whole-body insulin sensitivity as well as left ventricular ejection fraction improved [114].

Together these data suggest that a decrease in cardiac lipid content is not a prerequisite for improved left ventricular function. Several factors might explain this. Firstly, a similar mechanism like with IMCLs in endurance-trained athletes (have increased IMCLs but an increased fatty acid oxidative capacity) might play a role. It would be, therefore, interesting to study endurance-trained athletes to see whether they have, like in skeletal muscle, higher levels of intramyocardial TG than healthy controls and whether this is associated with increased myocardial fatty acid oxidation and is related with cardiac function. On the other hand, improved insulin sensitivity with lower glucose and lipid levels might account for the improvement in cardiac function in the above 2 studies. However, plasma lipid levels did not change in the 2 studies.

DISCUSSION

T2DM is a multifactorial disease in which genetic, environmental and lifestyle factors induce insulin resistance and impaired insulin secretion, ultimately leading to chronic hyperglycemia and its complications. Given the association between T2DM and obesity the recent focus of research has been the link between them. Vague already described a link between visceral adipose tissue, insulin resistance and T2DM in 1947 [115]. But it has not been until the start of the obesity and diabetic epidemic that further elaboration on his work started. This research revealed that adipose tissue is not merely a storage depot for TG but actively secretes a vast array of factors such as cytokines, metalloproteinases, and adipokines that can induce inflammation and insulin resistance [9].

In addition, with the advancement of radiological techniques it has become apparent that patients with insulin resistance and T2DM not only have a higher visceral to subcutaneous fat ratio as compared to healthy subjects but that TG are also stored in other organs called ectopic fat depositions, for example, in the liver, skeletal muscle, heart, and perhaps the pancreas [19, 116].

It has been proposed that genetic, environmental, and behavioral factors determine the response of adipose tissue to excess energy intake over energy expenditure. A physiologic response to this would be adipocyte hyperplasia, that is both an increase in adipocyte number with a small increase in adipocyte size. A pathologic response leads to adipocyte hypertrophy. Hypertrophic adipocytes become stressed because of hypoxia and nutritional overload of the endoplasmatic reticulum. Subsequently, stress and inflammatory pathways are activated. The adipocytes secrete cytokines and chemokines that interfere with the insulin signaling pathway. In addition, macrophages are attracted in the stromal vascular fraction of adipose tissue and sustain a chronic inflammatory response. Insulin-resistant hypertrophic adipocytes have increased lipolytic activity, together with the impaired ability of to take up FFAs, and consequently, a redirection of lipids towards nonadipose tissues (ectopic fat deposition) ensues [10].

If fatty acid beta-oxidation in the mitochondria cannot keep up with the increased supply of FFA to nonadipose tissues such as the liver, skeletal muscle, pancreas, and heart, accumulation of lipid intermediates like DAG and ceramides occurs. These lipid intermediates lead to activation of serine/threonine kinases that phosphorylate IRS molecules on serine residues. Serine phosphorylated IRS do not function properly, hence insulin signaling is impaired, and normal metabolic processes are disrupted.

Indeed, ectopic fat in the liver [67] and muscle [26] is positively correlated with insulin resistance and T2DM. Myocardial steatosis is associated with impaired diastolic function [104], and fat around the heart is associated with coronary artery disease [93] and related to whole-body insulin resistance [90]. The consequences of elevated fatty acids and/or lipid accumulation in the pancreas seem to be only in order when elevated glucose levels are already present [11].

The data in this paper show that substantial weight loss mobilizes ectopic fat stores in all organs and that this is associated with an improvement of the function of that organ. Thus, a reduction in hepatic TG was accompanied by a decline in fasting EGP [41, 73] and an improvement in the insulin suppressibility of EGP. A decrease in myocardial TG [112] and epicardial fat [97] were both associated with improved diastolic cardiac function. Finally, a decline in IMCLs leads to an improved insulin-stimulated glucose disposal [50].

It should be noted, however, that the amount of weight loss and/or the severity of caloric restriction are of influence on this positive effect and that there seems to be a tissuespecific reaction. For example, around eight kilograms weight loss following a 1200 kcal/day diet for 7 weeks led to a decrease in hepatic TG and improved insulin sensitivity of the liver but had no effect on insulin-stimulated glucose disposal or IMCLs in obese T2DM patients [41]. When obese women with a history of gestational diabetes were subdivided into groups with high and low liver fat, a similar weight loss led to greater loss of hepatic fat in the high liver fat group while both groups lost an equal amount of visceral and subcutaneous fat [72]. A VLCD for 8 weeks in moderately obese (BMI 33.6 ± 1.2 kg/m^2) patients with T2DM < 4 years duration decreased body weight by 15.3 ± 1.2 kg. Like the study with 8 kg weight loss by hypocaloric diet, this decreased IHL content and improved hepatic insulin sensitivity but had no effect on insulin-stimulated glucose disposal. Apparently, either a longer duration of severe caloric restriction or a greater weight loss is necessary to improve insulin-stimulated glucose disposal. We showed that a prolonged VLCD in more severely obese insulin-dependent T2DM patients leading to around 22 kg of weight reduction improved insulinstimulated glucose disposal and decreased IMCLs [50]. We recently corroborated these findings in a similar group of patients who underwent a 16-week VLCD (clamp data submitted for publication): ~27 kg weight loss improved both hepatic as well as peripheral insulin sensitivity. The largest reduction occurred in hepatic TG content (−85%), whereas IMCLs accumulation in the skeletal muscle decreased by 38%. The relative reduction in visceral fat was larger than the reduction in subcutaneous abdominal fat (−60% and −45% resp.) (accepted for publication in Obesity [98]).

The above studies suggest that hepatic TG content is the most easily mobilized, followed by visceral fat. The tissue-specific reaction to dietary interventions is also present when, vice versa, patients are subjected to high-fat feeding. Three days of a high-fat high-energy diet in health young males greatly increased hepatic TG stores but had no effect on myocardial TG [117].

Few studies have investigated the effect of exercise per se, that is, exercise without weight loss and/or caloric restriction. The effect of exercise varies in the different organs with respect to TG accumulation. In muscle, exercise can even increase IMCLs [47, 55, 56]. However, when this is accompanied with increased fatty acid oxidation this is positive and in accordance with the athlete's paradox. The latter refers to the fact that endurance-trained athletes have increased IMCLs but are very insulin sensitive [26]. In these athletes the IMCLs are a substrate source during exercise, and the high turnover rate prevents accumulation of lipid intermediates that have a negative effect on insulin signaling and can form lipid peroxides. In the sedentary state, when metabolic flexibility is low, IMCLs accumulate with the afore-mentioned deleterious effect on cellular processes [27]. Exercise alone can either increase [47, 55, 56] or decrease IMCLs [54] but does improve insulin sensitivity (Table 1). Apart from a decrease in lipid intermediates with increased fatty acid oxidation, an increase in capillary density and activation of AMPK with subsequently enhanced GLUT4 translocation might also be involved in the observed improvement in peripheral insulin sensitivity. Exercise in combination with diet also depletes hepatic TG content and improves hepatic insulin sensitivity [78], whereas it is undecided yet whether exercise alone depletes hepatic TG content [80, 81]. The underlying mechanism for depletion of IHL is probably not direct but via a decrease in factors produced by adipose [9] and skeletal muscle tissue. Exercise only reduces epicardial fat, but cardiac function was not measured in that study [99]. Exercise improved left ventricular ejection fraction after a 12-week exercise intervention in obese NGT and obese T2DM subjects. However, myocardial TG content in the T2DM patients were not affected, suggesting that either a decrease in myocardial TG content is not relevant for the improvement in cardiac function, or that we might be looking to the same paradox found in skeletal muscle (athletes paradox [26]).

Further research should focus both on the origins and processes involved in adipose tissue dysfunction as well as on the consequences of ectopic fat on the cellular level. One of the key questions is whether the impaired fatty acid oxidation is the cause or consequence of impaired mitochondrial function. Recent evidence points to the latter [38]. That might have consequences for therapy. In addition, the role of the gut microbiota in inducing inflammation and insulin resistance should be unraveled [118, 119]. Nevertheless, given the fact that subjects adopting a Western diet rapidly become insulin resistant [120], and that this diet both changes gut microbiota, inflammation, and adiposity (vice versa weight loss and healthy diets change these parameters for the better) strongly supports that excess dietary fat induces an inflammatory response and impairs mitochondrial function if lipid oxidation cannot keep up with increased FFA delivery. Diet and exercise are powerful tools in improving both ectopic fat deposition and the function of the organ in which the ectopic fat is deposited. Diet and lifestyle intervention, therefore, deserve more attention, both as preventive measure for obesity and T2DM as well as for the treatment of insulin resistance and T2DM.

When prescribing diet and exercise, it appears that exercise only has little effect on body weight and depletion of intracellular lipid stores but does improve skeletal muscle insulin sensitivity (most likely via increased contraction mediated GLUT-4 translocation) and perhaps cardiac function. With respect to diets only, moderate caloric restriction (>1200 kcal/day) and moderate weight loss (3–15 kg) only improve hepatic insulin sensitivity. More severe energy restriction (VLCD < 800 kcal/day) for a longer period (>8 weeks) and/or greater weight loss (>20 kg) are necessary to improve insulin-stimulated peripheral glucose uptake.

APPENDIX

A. Literature Search Details

A.1. Pubmed

("Diabetes Mellitus, Type 2" [Majr] OR ((diabetic [ti] OR "diabetes mellitus" [ti]) AND "type 2" [ti]) OR "obesity" [Majr] OR "insulin resistance" [Majr] OR "obesity" [ti] OR "insulin resistance" [ti] OR obese [ti] OR "Non-Insulin-Dependent Diabetes Mellitus" [ti] OR "Type 2 Diabetes Mellitus" [ti] OR "Stable Diabetes Mellitus" [ti] OR "Maturity-Onset Diabetes Mellitus" [ti] OR MODY [ti] OR NIDDM [ti] OR "Adult-Onset Diabetes Mellitus" [ti]) AND "Weight Loss" [Mesh] OR "weight reduction" [tw] OR "weight loss" [tw] OR "Diet" [mesh] OR diet [tw] OR diets [tw] OR "Carbohydrate-Restricted" [tw] OR "Fat-Restricted" [tw] OR "Caloric Restriction" [tw] OR "Caloric Restrictions" [tw] OR "low carb" [tw] OR "low fat" [tw] OR "low caloric" [tw] OR VLCD [tw] OR "low energy" [tw] OR "low calorie" [tw] OR "Diet Therapy" [Mesh] OR "Behavior Therapy" [Mesh] OR "behavior therapy" [tw] OR "behaviour therapy" [tw] OR "conditioning therapy" [tw] OR "behavior modification" [tw] OR "behavior modifications" [tw] OR "cognitive therapy" [tw] OR "behaviour modification" [tw] OR "behaviour modifications" [tw] OR "Exercise" [mesh] OR exercise [tw] OR "Sports" [Mesh] OR sports [tw] OR sport [tw] AND ("ectopic fat" [tw] OR "ectopic lipid" [tw] OR "ectopic lipids" [tw] OR "hepatic steatosis" [tw] OR "fatty liver" [mesh] OR "fatty liver" [tw] OR NAFLD [tw] OR NASH [tiab] OR "nonalcoholic steatohepatitis" [tw] OR "nonalcoholic steatohepatitis" [tw] OR "liver steatosis" [tw] OR "hepatic triglyceride" [tw] OR "hepatic triglycerides" [tw] OR "liver triglyceride" [tw] OR "hepatic triglycerides" [tw] OR "Myocardial triglyceride" [tw] OR "Myocardial triglycerides" [tw] OR "Epicardial fat" [tw] OR "pericardial fat" [tw] OR "mediastinal fat" [tw] OR "intrathoracic fat" [tw] OR "intramyocellular triglyceride" [tw] OR "intramyocellular triglycerides" [tw] OR "intramyocellular lipid" [tw] OR "intramyocellular lipid" [tw] OR "intramyocellular fat" [tw] OR IMCLs [tw]).

A.2. EMBASE (OVID-Version)

(*non insulin dependent diabetes mellitus/OR ((diabetic OR "diabetes mellitus") AND "type 2").ti OR exp *obesity/OR *insulin resistance/OR obesity.ti OR insulin resistance.ti OR obese.ti OR Non-Insulin-Dependent Diabetes Mellitus.ti OR "Type 2 Diabetes".ti OR "Stable Diabetes Mellitus".ti OR "Maturity-Onset Diabetes Mellitus".ti OR MODY.ti OR NIDDM.ti OR "Adult-Onset Diabetes Mellitus".ti) AND (weight reduction/OR weight reduction.mp OR weight loss.mp OR exp diet/OR diet.mp OR diets.mp

OR exp diet therapy/OR exp dietary intake/OR Carbohydrate Restrict*. mp OR Fat Restrict*.mp OR Caloric Restrict*.mp OR low carb.mp OR low fat.mp OR low caloric.mp OR VLCD.mp OR low energy.mp OR low calorie.mp OR behavior therapy/OR behavior therapy.mp OR behaviour therapy.mp OR conditioning therapy.mp OR behavior modification/OR behavior modification*.mp OR cognitive therapy/OR psychotherapy/OR cognitive therapy.mp OR behaviour modification*.mp OR exp exercise/ OR exercise.mp OR exp Sport/OR exp physical activity/OR sports.mp OR sport.mp AND (ectopic fat*.mp OR ectopic lipid*.mp OR exp fatty liver/ OR hepatic steatosis.mp OR fatty liver.mp OR NAFLD.mp OR NASH. mp OR nonalcoholic steatohepatitis.mp OR nonalcoholic steatohepatitis. mp OR liver steatosis.mp OR hepatic triglycerid*.mp OR liver triglycerid* OR myocardial triglycerid*.mp OR epicardial fat.mp OR pericardial fat*. mp OR mediastinal fat*.mp OR intrathoracic fat*.mp OR intramyocellular triglycerid*.mp OR intramyocellular lipid*.mp OR intramyocellular fat*. mp OR IMCLs.mp).

A.3. Web of Science

TI = (non insulin dependent diabetes mellitus OR ((diabetic OR "diabetes mellitus") AND "type 2") OR obesity OR insulin resistance OR insulin resistance OR obese OR Non-Insulin-Dependent Diabetes Mellitus OR "Type 2 Diabetes" OR "Stable Diabetes Mellitus" OR "Maturity-Onset Diabetes Mellitus" OR MODY OR NIDDM OR "Adult-Onset Diabetes Mellitus") AND TS = (weight reduction OR weight loss OR diet OR diets OR dietary intake OR Carbohydrate Restrict* OR Fat Restrict* OR Caloric Restrict* OR low carb OR low fat OR low caloric OR VLCD OR low energy OR low calorie OR behavior therapy OR behaviour therapy OR conditioning therapy OR behavior modification* OR cognitive therapy OR psychotherapy OR behaviour modification* OR exercise OR physical activity OR sports OR sport) AND TS = ("ectopic fat*" OR "ectopic lipid*" OR "fatty liver" OR "hepatic steatosis" OR NAFLD OR NASH OR "nonalcoholic steatohepatitis" OR "nonalcoholic steatohepatitis" OR "liver steatosis" OR "hepatic triglycerid*" OR "liver triglycerid*" OR "myocardial triglycerid*" OR "epicardial fat" OR "pericardial fat*" OR "mediastinal fat*" OR "intrathoracic fat*" OR "intramyocellular triglycerid*" OR "intramyocellular lipid*" OR "intramyocellular fat*" OR IMCLs).

A.4. Cochrane Library

ID Search#1 MeSH descriptor Diabetes Mellitus, Type 2 explode all trees#2 MeSH descriptor Obesity explode all trees#3 MeSH descriptor Insulin Resistance explode all trees#4 (diabet* AND "type 2") OR obesity OR insulin resistance OR obese OR "Non-Insulin-Dependent Diabetes Mellitus" OR "Type 2 Diabetes Mellitus" OR "Stable Diabetes Mellitus" OR "Maturity-Onset Diabetes Mellitus" OR MODY OR NIDDM OR "Adult-Onset Diabetes Mellitus":ti#5 (#1 OR #2 OR #3 OR #4)#6 MeSH descriptor Weight Loss explode all trees#7 MeSH descriptor Diet explode all trees#8 MeSH descriptor Diet Therapy explode all trees#9 MeSH descriptor Behavior Therapy explode all trees#10 MeSH descriptor Exercise explode all trees#11 MeSH descriptor Sports explode all trees#12 (("anti obesity" OR "anti-obesity" OR antiobesity OR "weight-loss" OR "weight loss") AND "weight reduction" OR "weight loss" OR diet OR diets OR "Carbohydrate-Restricted" OR "Fat-Restricted" OR "Caloric Restriction" OR "Caloric Restrictions" OR "low carb" OR "low fat" OR "low caloric" OR VLCD OR "low energy" OR "low calorie" OR "behavior therapy" OR "behaviour therapy" OR "conditioning therapy" OR "behavior modification" OR "behavior modifications" OR "cognitive therapy" OR "behaviour modification" OR "behaviour modifications" OR exercise OR sports OR sport *:ti,ab,kw#13 (#6 OR #7 OR #8 OR #9 OR #10 OR #11 OR #12)#14 MeSH descriptor Fatty Liver explode all trees#15 ("ectopic fat" OR "ectopic lipid" OR "ectopic lipids" OR "hepatic steatosis" OR "fatty liver" OR NAFLD OR NASH OR "nonalcoholic steatohepatitis" OR "nonalcoholic steatohepatitis" OR "liver steatosis" OR "hepatic triglyceride" OR "hepatic triglycerides" OR "liver triglyceride" OR "hepatic triglycerides" OR "Myocardial triglyceride" OR "Myocardial triglycerides" OR "Epicardial fat" OR "pericardial fat" OR "mediastinal fat" OR "intrathoracic fat" OR "intramyocellular triglyceride" OR "intramyocellular triglycerides" OR "intramyocellular lipid" OR "intramyocellular lipid" OR "intramyocellular fat" OR IMCLs): ti,ab,kw#16 (#14 OR #15)#17 (#5 AND #13 AND #16).

REFERENCES

1. "Obesity and overweight," WHO Report No.: fact sheet no 311, 2012, http://www.who.int/mediacentre/factsheets/fs311/en/index.html.
2. F. X. Pi-Sunyer, "Medical hazards of obesity," Annals of Internal Medicine, vol. 119, no. 7, pp. 655–660, 1993.
3. S. Z. Yanovski, "Overweight, obesity, and health risk: National Task Force on the Prevention and Treatment of Obesity," Archives of Internal Medicine, vol. 160, no. 7, pp. 898–904, 2000.
4. A. E. Field, E. H. Coakley, A. Must et al., "Impact of overweight on the risk of developing common chronic diseases during a 10-year period," Archives of Internal Medicine, vol. 161, no. 13, pp. 1581–1586, 2001.
5. W. C. Willett, W. H. Dietz, and G. A. Colditz, "Guidelines for healthy weight," New England Journal of Medicine, vol. 341, no. 6, pp. 427–434, 1999.
6. J. M. Chan, E. B. Rimm, G. A. Colditz, M. J. Stampfer, and W. C. Willett, "Obesity, fat distribution, and weight gain as risk factors for clinical diabetes in men," Diabetes Care, vol. 17, no. 9, pp. 961–969, 1994.·
7. G. A. Colditz, W. C. Willett, A. Rotnitzky, and J. E. Manson, "Weight gain as a risk factor for clinical diabetes mellitus in women," Annals of Internal Medicine, vol. 122, no. 7, pp. 481–486, 1995.
8. R. P. Wildman, P. Muntner, K. Reynolds et al., "The obese without cardiometabolic risk factor clustering and the normal weight with cardiometabolic risk factor clustering: prevalence and correlates of 2 phenotypes among the US population (NHANES 1999-2004)," Archives of Internal Medicine, vol. 168, no. 15, pp. 1617–1624, 2008.
9. I. M. Jazet, H. Pijl, and A. E. Meinders, "Adipose tissue as an endocrine organ: impact on insulin resistance," Netherlands Journal of Medicine, vol. 61, no. 6, pp. 194–212, 2003.
10. M. Blüher, "Adipose tissue dysfunction in obesity," Experimental and Clinical Endocrinology and Diabetes, vol. 117, no. 6, pp. 241–250, 2009.
11. D. H. Van Raalte, N. J. Van Der Zijl, and M. Diamant, "Pancreatic steatosis in humans: cause or marker of lipotoxicity?" Current Opinion

in Clinical Nutrition and Metabolic Care, vol. 13, no. 4, pp. 478–485, 2010.

12. N. J. Van Der Zijl, G. H. Goossens, C. C. M. Moors et al., "Ectopic fat storage in the pancreas, liver, and abdominal fat depots: impact on β-cell function in individuals with impaired glucose metabolism," Journal of Clinical Endocrinology and Metabolism, vol. 96, no. 2, pp. 459–467, 2011.
13. E. L. Lim, K. G. Hollingsworth, B. S. Aribisala, M. J. Chen, J. C. Mathers, and R. Taylor, "Reversal of type 2 diabetes: normalisation of beta cell function in association with decreased pancreas and liver triacylglycerol," Diabetologia, vol. 54, no. 10, pp. 2506–2514, 2011.
14. M. Jernås, J. Palming, K. Sjöholm et al., "Separation of human adipocytes by size: hypertrophic fat cells display distinct gene expression," The FASEB Journal, vol. 20, no. 9, pp. 1540–1542, 2006.
15. U. Özcan, Q. Cao, E. Yilmaz et al., "Endoplasmic reticulum stress links obesity, insulin action, and type 2 diabetes," Science, vol. 306, no. 5695, pp. 457–461, 2004.
16. M. F. Gregor and G. S. Hotamisligil, "Adipocyte stress: the endoplasmic reticulum and metabolic disease," Journal of Lipid Research, vol. 48, no. 9, pp. 1905–1914, 2007.
17. J. F. C. Glatz, A. Bonen, and J. J. F. P. Luiken, "Exercise and insulin increase muscle fatty acid uptake by recruiting putative fatty acid transporters to the sarcolemma," Current Opinion in Clinical Nutrition and Metabolic Care, vol. 5, no. 4, pp. 365–370, 2002.
18. K. Morino, K. F. Petersen, and G. I. Shulman, "Molecular mechanisms of insulin resistance in humans and their potential links with mitochondrial dysfunction," Diabetes, vol. 55, no. 2, pp. S9–S15, 2006.
19. J. Szendroedi and M. Roden, "Ectopic lipids and organ function," Current Opinion in Lipidology, vol. 20, no. 1, pp. 50–56, 2009.
20. V. T. Samuel, Z. X. Liu, X. Qu et al., "Mechanism of hepatic insulin resistance in non-alcoholic fatty liver disease," Journal of Biological Chemistry, vol. 279, no. 31, pp. 32345–32353, 2004.
21. M. Krssak, K. Falk Petersen, A. Dresner et al., "Intramyocellular lipid concentrations are correlated with insulin sensitivity in humans: a 1H NMR spectroscopy study," Diabetologia, vol. 42, no. 1, pp. 113–116, 1999.
22. G. Perseghin, P. Scifo, F. De Cobelli et al., "Intramyocellular

triglyceride content is a determinant of in vivo insulin resistance in humans: a 1H-13C nuclear magnetic resonance spectroscopy assessment in offspring of type 2 diabetic parents," Diabetes, vol. 48, no. 8, pp. 1600–1606, 1999.

23. D. A. Pan, S. Lillioja, A. D. Kriketos et al., "Skeletal muscle triglyceride levels are inversely related to insulin action," Diabetes, vol. 46, no. 6, pp. 983–988, 1997.
24. S. Jacob, J. Machann, K. Rett et al., "Association of increased intramyocellular lipid content with insulin resistance in lean nondiabetic offspring of type 2 diabetic subjects," Diabetes, vol. 48, no. 5, pp. 1113–1119, 1999.
25. D. I. W. Phillips, S. Caddy, V. Ilic et al., "Intramuscular triglyceride and muscle insulin sensitivity: evidence for a relationship in nondiabetic subjects," Metabolism, vol. 45, no. 8, pp. 947–950, 1996.
26. B. H. Goodpaster, J. He, S. Watkins, and D. E. Kelley, "Skeletal muscle lipid content and insulin resistance: evidence for a paradox in endurance-trained athletes," Journal of Clinical Endocrinology and Metabolism, vol. 86, no. 12, pp. 5755–5761, 2001.
27. L. J. C. Van Loon and B. H. Goodpaster, "Increased intramuscular lipid storage in the insulin-resistant and endurance-trained state," Pflugers Archiv European Journal of Physiology, vol. 451, no. 5, pp. 606–616, 2006.
28. D. E. Kelley, B. Goodpaster, R. R. Wing, and J. A. Simoneau, "Skeletal muscle fatty acid metabolism in association with insulin resistance, obesity, and weight loss," American Journal of Physiology, vol. 277, no. 6, pp. E1130–E1141, 1999.
29. J. A. Simoneau and D. E. Kelley, "Altered glycolytic and oxidative capacities of skeletal muscle contribute to insulin resistance in NIDDM," Journal of Applied Physiology, vol. 83, no. 1, pp. 166–171, 1997.
30. J. A. Simoneau, J. H. Veerkamp, L. P. Turcotte, and D. E. Kelley, "Markers of capacity to utilize fatty acids in human skeletal muscle: relation to insulin resistance and obesity and effects of weight loss," FASEB Journal, vol. 13, no. 14, pp. 2051–2060, 1999.
31. K. F. Petersen, S. Dufour, D. Befroy, R. Garcia, and G. I. Shulman, "Impaired mitochondrial activity in the insulin-resistant offspring of patients with type 2 diabetes," New England Journal of Medicine, vol.

350, no. 7, pp. 664–671, 2004.

32. K. Morino, K. F. Petersen, S. Dufour et al., "Reduced mitochondrial density and increased IRS-1 serine phosphorylation in muscle of insulin-resistant offspring of type 2 diabetic parents," Journal of Clinical Investigation, vol. 115, no. 12, pp. 3587–3593, 2005.
33. K. F. Petersen, S. Dufour, and G. I. Shulman, "Decreased insulin-stimulated ATP synthesis and phosphate transport in muscle of insulin-resistant offspring of type 2 diabetic parents," PLoS Medicine, vol. 2, no. 9, article e233, 2005.
34. D. E. Kelley, J. He, E. V. Menshikova, and V. B. Ritov, "Dysfunction of mitochondria in human skeletal muscle in type 2 diabetes," Diabetes, vol. 51, no. 10, pp. 2944–2950, 2002.
35. J. Szendroedi, A. I. Schmid, M. Chmelik et al., "Muscle mitochondrial ATP synthesis and glucose transport/phosphorylation in type 2 diabetes," PLoS Medicine, vol. 4, no. 5, article e154, 2007.
36. E. Phielix, V. B. Schrauwen-Hinderling, M. Mensink et al., "Lower intrinsic ADP-stimulated mitochondrial respiration underlies in vivo mitochondrial dysfunction in muscle of male type 2 diabetic patients," Diabetes, vol. 57, no. 11, pp. 2943–2949, 2008.
37. V. B. Schrauwen-Hinderling, M. E. Kooi, M. K. C. Hesselink et al., "Impaired in vivo mitochondrial function but similar intramyocellular lipid content in patients with type 2 diabetes mellitus and BMI-matched control subjects," Diabetologia, vol. 50, no. 1, pp. 113–120, 2007.
38. P. Schrauwen, V. Schrauwen-Hinderling, J. Hoeks, and M. K. C. Hesselink, "Mitochondrial dysfunction and lipotoxicity," Biochimica et Biophysica Acta, vol. 1801, no. 3, pp. 266–271, 2010.
39. A. V. Greco, G. Mingrone, A. Giancaterini et al., "Insulin resistance in morbid obesity: reversal with intramyocellular fat depletion," Diabetes, vol. 51, no. 1, pp. 144–151, 2002.
40. P. Malenfant, D. R. Joanisse, R. Thériault, B. H. Goodpaster, D. E. Kelley, and J.-A. Simoneau, "Fat content in individual muscle fibers of lean and obese subjects," International Journal of Obesity, vol. 25, no. 9, pp. 1316–1321, 2001.
41. K. F. Petersen, S. Dufour, D. Befroy, M. Lehrke, R. E. Hendler, and G. I. Shulman, "Reversal of nonalcoholic hepatic steatosis, hepatic insulin resistance, and hyperglycemia by moderate weight reduction in patients with type 2 diabetes," Diabetes, vol. 54, no. 3, pp. 603–608,

2005.

42. F. Sato, Y. Tamura, H. Watada et al., "Brief report: effects of diet-induced moderate weight reduction on intrahepatic and intramyocellular triglycerides and glucose metabolism in obese subjects," Journal of Clinical Endocrinology and Metabolism, vol. 92, no. 8, pp. 3326–3329, 2007.
43. Y. Tamura, Y. Tanaka, F. Sato et al., "Effects of diet and exercise on muscle and liver intracellular lipid contents and insulin sensitivity in type 2 diabetic patients," Journal of Clinical Endocrinology and Metabolism, vol. 90, no. 6, pp. 3191–3196, 2005.
44. G. Mingrone, G. Rosa, P. Di Rocco et al., "Skeletal muscle triglycerides lowering is associated with net improvement of insulin sensitivity, TNF-α reduction and GLUT4 expression enhancement," International Journal of Obesity, vol. 26, no. 9, pp. 1165–1172, 2002.
45. D. E. Larson-Meyer, L. K. Heilbronn, L. M. Redman et al., "Effect of calorie restriction with or without exercise on insulin sensitivity, β-cell function, fat cell size, and ectopic lipid in overweight subjects," Diabetes Care, vol. 29, no. 6, pp. 1337–1344, 2006.
46. P. Malenfant, A. Tremblay, E. Doucet, P. Imbeault, J.-A. Simoneau, and D. R. Joanisse, "Elevated intramyocellular lipid concentration in obese subjects is not reduced after diet and exercise training," American Journal of Physiology, vol. 280, no. 4 43-4, pp. E632–E639, 2001.
47. J. J. Dubé, F. Amati, F. G. S. Toledo et al., "Effects of weight loss and exercise on insulin resistance, and intramyocellular triacylglycerol, diacylglycerol and ceramide," Diabetologia, vol. 54, no. 5, pp. 1147–1156, 2011.
48. S. B. Haugaard, A. Vaag, C. E. Høy, and S. Madsbad, "Desaturation of skeletal muscle structural and depot lipids in obese individuals during a very-low-calorie diet intervention," Obesity, vol. 15, no. 1, pp. 117–125, 2007.
49. C. Lara-Castro, B. R. Newcomer, J. Rowell et al., "Effects of short-term very low-calorie diet on intramyocellular lipid and insulin sensitivity in nondiabetic and type 2 diabetic subjects," Metabolism: Clinical and Experimental, vol. 57, no. 1, pp. 1–8, 2008.
50. I. M. Jazet, G. Schaart, A. Gastaldelli et al., "Loss of 50% of excess weight using a very low energy diet improves insulin-stimulated glucose disposal and skeletal muscle insulin signalling in obese

insulin-treated type 2 diabetic patients," Diabetologia, vol. 51, no. 2, pp. 309–319, 2008.

51. F. G. S. Toledo, E. V. Menshikova, K. Azuma et al., "Mitochondrial capacity in skeletal muscle is not stimulated by weight loss despite increases in insulin action and decreases in intramyocellular lipid content," Diabetes, vol. 57, no. 4, pp. 987–994, 2008.
52. F. G. S. Toledo, E. V. Menshikova, V. B. Ritov et al., "Effects of physical activity and weight loss on skeletal muscle mitochondria and relationship with glucose control in type 2 diabetes," Diabetes, vol. 56, no. 8, pp. 2142–2147, 2007.
53. J. He, B. H. Goodpaster, and D. E. Kelley, "Effects of weight loss and physical activity on muscle lipid content and droplet size," Obesity Research, vol. 12, no. 5, pp. 761–769, 2004.
54. T. P. J. Solomon, S. N. Sistrun, R. K. Krishnan et al., "Exercise and diet enhance fat oxidation and reduce insulin resistance in older obese adults," Journal of Applied Physiology, vol. 104, no. 5, pp. 1313–1319, 2008.
55. J. J. Dubé, F. Amati, M. Stefanovic-Racic, F. G. S. Toledo, S. E. Sauers, and B. H. Goodpaster, "Exercise-induced alterations in intramyocellular lipids and insulin resistance: the athlete's paradox revisited," American Journal of Physiology, vol. 294, no. 5, pp. E882–E888, 2008.
56. R. C. R. Meex, V. B. Schrauwen-Hinderling, E. Moonen-Kornips et al., "Restoration of muscle mitochondrial function and metabolic flexibility in type 2 diabetes by exercise training is paralleled by increased myocellular fat storage and improved insulin sensitivity," Diabetes, vol. 59, no. 3, pp. 572–579, 2010.
57. M. Lazo and J. M. Clark, "The epidemiology of nonalcoholic fatty liver disease: a global perspective," Seminars in Liver Disease, vol. 28, no. 4, pp. 339–350, 2008.
58. T. Andersen and C. Gluud, "Liver morphology in morbid obesity: a literature study," International Journal of Obesity, vol. 8, no. 2, pp. 97–106, 1984.
59. S. Bellentani, G. Saccoccio, F. Masutti et al., "Prevalence of and risk factors for hepatic steatosis in northern Italy," Annals of Internal Medicine, vol. 132, no. 2, pp. 112–117, 2000.
60. E. Fabbrini, S. Sullivan, and S. Klein, "Obesity and nonalcoholic fatty liver disease: biochemical, metabolic, and clinical

implications," Hepatology, vol. 51, no. 2, pp. 679–689, 2010.

61. L. S. Szczepaniak, P. Nurenberg, D. Leonard et al., "Magnetic resonance spectroscopy to measure hepatic triglyceride content: prevalence of hepatic steatosis in the general population," American Journal of Physiology, vol. 288, no. 2, pp. E462–E468, 2005.
62. K. F. Petersen, S. Dufour, J. Feng et al., "Increased prevalence of insulin resistance and nonalcoholic fatty liver disease in Asian-Indian men," Proceedings of the National Academy of Sciences of the United States of America, vol. 103, no. 48, pp. 18273–18277, 2006.
63. D. E. Kleiner, E. M. Brunt, M. Van Natta et al., "Design and validation of a histological scoring system for nonalcoholic fatty liver disease," Hepatology, vol. 41, no. 6, pp. 1313–1321, 2005.
64. K. M. Korenblat, E. Fabbrini, B. S. Mohammed, and S. Klein, "Liver, muscle, and adipose tissue insulin action is directly related to intrahepatic triglyceride content in obese subjects," Gastroenterology, vol. 134, no. 5, pp. 1369–1375, 2008.
65. E. Bugianesi, A. Gastaldelli, E. Vanni et al., "Insulin resistance in non-diabetic patients with non-alcoholic fatty liver disease: sites and mechanisms," Diabetologia, vol. 48, no. 4, pp. 634–642, 2005.
66. A. J. Sanyal, C. Campbell-Sargent, F. Mirshahi et al., "Nonalcoholic steatohepatitis: association of insulin resistance and mitochondrial abnormalities," Gastroenterology, vol. 120, no. 5, pp. 1183–1192, 2001.
67. A. Seppälä-Lindroos, S. Vehkavaara, A.-M. Häkkinen et al., "Fat accumulation in the liver is associated with defects in insulin suppression of glucose production and serum free fatty acids independent of obesity in normal men," Journal of Clinical Endocrinology and Metabolism, vol. 87, no. 7, pp. 3023–3028, 2002.
68. A. Gastaldelli, M. Kozakova, K. Höjlund et al., "Fatty liver is associated with insulin resistance, risk of coronary heart disease, and early atherosclerosis in a large European population," Hepatology, vol. 49, no. 5, pp. 1537–1544, 2009.
69. S. L. Colles, J. B. Dixon, P. Marks, B. J. Strauss, and P. E. O'Brien, "Preoperative weight loss with a very-low-energy diet: quantitation of changes in liver and abdominal fat by serial imaging," American Journal of Clinical Nutrition, vol. 84, no. 2, pp. 304–311, 2006.
70. C. Thamer, J. Machann, N. Stefan et al., "High visceral fat mass and high

liver fat are associated with resistance to lifestyle intervention," Obesity, vol. 15, no. 2, pp. 531–538, 2007.

71. E. L. Thomas, A. E. Brynes, G. Hamilton et al., "Effect of nutritional counselling on hepatic, muscle and adipose tissue fat content and distribution in non-alcoholic fatty liver disease," World Journal of Gastroenterology, vol. 12, no. 36, pp. 5813–5819, 2006.
72. M. Tiikkainen, R. Bergholm, S. Vehkavaara et al., "Effects of identical weight loss on body composition and features of insulin resistance in obese women with high and low liver fat content," Diabetes, vol. 52, no. 3, pp. 701–707, 2003.
73. A. P. M. Viljanen, P. Iozzo, R. Borra et al., "Effect of weight loss on liver free fatty acid uptake and hepatic insulin resistance," Journal of Clinical Endocrinology and Metabolism, vol. 94, no. 1, pp. 50–55, 2009.
74. M. A. Huang, J. K. Greenson, C. Chao et al., "One-year intense nutritional counseling results in histological improvement in patients with nonalcoholic steatohepatitis: a pilot study," American Journal of Gastroenterology, vol. 100, no. 5, pp. 1072–1081, 2005.
75. T. Ueno, H. Sugawara, K. Sujaku et al., "Therapeutic effects of restricted diet and exercise in obese patients with fatty liver," Journal of Hepatology, vol. 27, no. 1, pp. 103–107, 1997.
76. T. Andersen, C. Gluud, M. B. Franzmann, and P. Christoffersen, "Hepatic effects of dietary weight loss in morbidly obese subjects," Journal of Hepatology, vol. 12, no. 2, pp. 224–229, 1991.
77. N. A. Spassiani and J. L. Kuk, "Exercise and the fatty liver," Applied Physiology, Nutrition and Metabolism, vol. 33, no. 4, pp. 802–807, 2008.
78. F. Magkos, "Exercise and fat accumulation in the human liver," Current Opinion in Lipidology, vol. 21, no. 6, pp. 507–517, 2010.
79. K. Shah, A. Stufflebam, T. N. Hilton, D. R. Sinacore, S. Klein, and D. T. Villareal, "Diet and exercise interventions reduce intrahepatic fat content and improve insulin sensitivity in obese older adults," Obesity, vol. 17, no. 12, pp. 2162–2168, 2009.
80. F. Shojaee-Moradie, K. C. Baynes, C. Pentecost et al., "Exercise training reduces fatty acid availability and improves the insulin sensitivity of glucose metabolism," Diabetologia, vol. 50, no. 2, pp. 404–413, 2007.·
81. N. A. Johnson, T. Sachinwalla, D. W. Walton et al., "Aerobic exercise

training reduces hepatic and visceral lipids in obese individuals without weight loss," Hepatology, vol. 50, no. 4, pp. 1105–1112, 2009.

82. C. A. Slentz, L. A. Bateman, L. H. Willis et al., "Effects of aerobic vs. resistance training on visceral and liver fat stores, liver enzymes, and insulin resistance by HOMA in overweight adults from STRRIDE AT/RT," American Journal of Physiology, vol. 301, no. 5, pp. E1033–E1039, 2011.
83. H. S. Sacks and J. N. Fain, "Human epicardial adipose tissue: a review," American Heart Journal, vol. 153, no. 6, pp. 907–917, 2007.
84. S. W. Rabkin, "Epicardial fat:properties, function and relationship to obesity," Obesity Reviews, vol. 8, no. 3, pp. 253–261, 2007.
85. J. M. Marchington, C. A. Mattacks, and C. M. Pond, "Adipose tissue in the mammalian heart and pericardium: structure, foetal development and biochemical properties," Comparative Biochemistry and Physiology B, vol. 94, no. 2, pp. 225–232, 1989.
86. J. M. Marchington and C. M. Pond, "Site-specific properties of pericardial and epicardial adipose tissue: the effects of insulin and high-fat feeding on lipogenesis and the incorporation of fatty acids in vitro," International Journal of Obesity, vol. 14, no. 12, pp. 1013–1022, 1990.
87. T. Mazurek, L. Zhang, A. Zalewski et al., "Human epicardial adipose tissue is a source of inflammatory mediators," Circulation, vol. 108, no. 20, pp. 2460–2466, 2003.
88. A. R. Baker, A. L. Harte, N. Howell et al., "Epicardial adipose tissue as a source of nuclear factor-κB and c-Jun N-terminal kinase mediated inflammation in patients with coronary artery disease," Journal of Clinical Endocrinology and Metabolism, vol. 94, no. 1, pp. 261–267, 2009.
89. S. G. Ahn, H. S. Lim, D. Y. Joe et al., "Relationship of epicardial adipose tissue by echocardiography to coronary artery disease," Heart, vol. 94, no. 3, article e7, 2008.
90. M. Greif, A. Becker, F. Von Ziegler et al., "Pericardial adipose tissue determined by dual source CT is a risk factor for coronary atherosclerosis," Arteriosclerosis, Thrombosis, and Vascular Biology, vol. 29, no. 5, pp. 781–786, 2009.
91. J. W. Jeong, H. J. Myung, H. Y. Kyeong et al., "Echocardiographic epicardial fat thickness and coronary artery disease," Circulation

Journal, vol. 71, no. 4, pp. 536–539, 2007.

92. G. A. Rosito, J. M. Massaro, U. Hoffmann et al., "Pericardial fat, visceral abdominal fat, cardiovascular disease risk factors, and vascular calcification in a community-based sample the framingham heart study," Circulation, vol. 117, no. 5, pp. 605–613, 2008.
93. G. Iacobellis and F. Leonetti, "Epicardial adipose tissue and insulin resistance in obese subjects," Journal of Clinical Endocrinology and Metabolism, vol. 90, no. 11, pp. 6300–6302, 2005.
94. C. P. Wang, H. L. Hsu, W. C. Hung et al., "Increased epicardial adipose tissue (EAT) volume in type 2 diabetes mellitus and association with metabolic syndrome and severity of coronary atherosclerosis," Clinical Endocrinology, vol. 70, no. 6, pp. 876–882, 2009.
95. T. D. Wang, W. J. Lee, F. Y. Shih et al., "Relations of epicardial adipose tissue measured by multidetector computed tomography to components of the metabolic syndrome are region-specific and independent of anthropometric indexes and intraabdominal visceral fat," Journal of Clinical Endocrinology and Metabolism, vol. 94, no. 2, pp. 662–669, 2009.
96. M. K. Kim, K. Tanaka, M. J. Kim et al., "Comparison of epicardial, abdominal and regional fat compartments in response to weight loss," Nutrition, Metabolism and Cardiovascular Diseases, vol. 19, no. 11, pp. 760–766, 2009.
97. G. Iacobellis, N. Singh, S. Wharton, and A. M. Sharma, "Substantial changes in epicardial fat thickness after weight loss in severely obese subjects," Obesity, vol. 16, no. 7, pp. 1693–1697, 2008.
98. M. Snel, J. T. Jonker, S. Hammer et al. et al., "Long-Term Beneficial Effect of a 16-Week Very Low Calorie Diet on Pericardial Fat in Obese Type 2 Diabetes Mellitus Patients," Obesity. In press.
99. M. K. Kim, T. Tomita, M. J. Kim, H. Sasai, S. Maeda, and K. Tanaka, "Aerobic exercise training reduces epicardial fat in obese men," Journal of Applied Physiology, vol. 106, no. 1, pp. 5–11, 2009.
100. T. E. Brinkley, J. Ding, J. J. Carr, and B. J. Nicklas, "Pericardial fat loss in postmenopausal women under conditions of equal energy deficit," Medicine & Science in Sports & Exercise, vol. 43, no. 5, pp. 808–814, 2011.
101. R. W. Van Der Meer, J. Doornbos, S. Kozerke et al., "Metabolic imaging of myocardial triglyceride content: reproducibility

of 1H MR spectroscopy with respiratory navigator gating in volunteers," Radiology, vol. 245, no. 1, pp. 251–257, 2007.

102. P. Iozzo, R. Lautamaki, R. Borra et al., "Contribution of glucose tolerance and gender to cardiac adiposity," Journal of Clinical Endocrinology and Metabolism, vol. 94, no. 11, pp. 4472–4482, 2009.
103. J. M. McGavock, I. Lingvay, I. Zib et al., "Cardiac steatosis in diabetes mellitus: a 1H-magnetic resonance spectroscopy study," Circulation, vol. 116, no. 10, pp. 1170–1175, 2007.
104. L. J. Rijzewijk, R. W. van der Meer, J. W. A. Smit et al., "Myocardial steatosis is an independent predictor of diastolic dysfunction in type 2 diabetes mellitus," Journal of the American College of Cardiology, vol. 52, no. 22, pp. 1793–1799, 2008.
105. Y. T. Zhou, P. Grayburn, A. Karim et al., "Lipotoxic heart disease in obese rats: implications for human obesity," Proceedings of the National Academy of Sciences of the United States of America, vol. 97, no. 4, pp. 1784–1789, 2000.
106. C. Christoffersen, E. Bollano, M. L. S. Lindegaard et al., "Cardiac lipid accumulation associated with diastolic dysfunction in obese mice," Endocrinology, vol. 144, no. 8, pp. 3483–3490, 2003.
107. L. R. Peterson, P. Herrero, K. B. Schechtman et al., "Effect of obesity and insulin resistance on myocardial substrate metabolism and efficiency in young women," Circulation, vol. 109, no. 18, pp. 2191–2196, 2004.
108. L. J. Rijzewijk, R. W. van der Meer, H. J. Lamb et al., "Altered myocardial substrate metabolism and decreased diastolic function in nonischemic human diabetic cardiomyopathy: studies with cardiac positron emission tomography and magnetic resonance imaging," Journal of the American College of Cardiology, vol. 54, no. 16, pp. 1524–1532, 2009.
109. S. Hammer, R. W. Van Der Meer, H. J. Lamb et al., "Short-term flexibility of myocardial triglycerides and diastolic function in patients with type 2 diabetes mellitus," American Journal of Physiology, vol. 295, no. 3, pp. E714–E718, 2008.
110. S. Hammer, R. W. Van Der Meer, H. J. Lamb et al., "Progressive caloric restriction induces dose-dependent changes in myocardial triglyceride content and diastolic function in healthy men," Journal of Clinical Endocrinology and Metabolism, vol. 93, no. 2, pp. 497–503, 2008.
111. A. P.M. Viljanen, A. Karmi, R. Borra et al., "Effect of caloric restriction

on myocardial fatty acid uptake, left ventricular mass, and cardiac work in obese adults," American Journal of Cardiology, vol. 103, no. 12, pp. 1721–1726, 2009.

112. S. Hammer, M. Snel, H. J. Lamb et al., "Prolonged caloric restriction in obese patients with type 2 diabetes mellitus decreases myocardial triglyceride content and improves myocardial function," Journal of the American College of Cardiology, vol. 52, no. 12, pp. 1006–1012, 2008.
113. V. B. Schrauwen-Hinderling, M. K. C. Hesselink, R. Meex et al., "Improved ejection fraction after exercise training in obesity is accompanied by reduced cardiac lipid content," Journal of Clinical Endocrinology and Metabolism, vol. 95, no. 4, pp. 1932–1938, 2010.
114. V. B. Schrauwen-Hinderling, R. C. R. Meex, M. K. C. Hesselink et al., "Cardiac lipid content is unresponsive to a physical activity training intervention in type 2 diabetic patients, despite improved ejection fraction," Cardiovascular Diabetology, vol. 10, article 47, 2011.
115. J. Vague, "La differenciation sexuelle: facteur determinant des formes: de l'obesite," La Presse Médicale, vol. 55, pp. 339–340, 1947.
116. E. Ravussin and S. R. Smith, "Increased fat intake, impaired fat oxidation, and failure of fat cell proliferation result in ectopic fat storage, insulin resistance, and type 2 diabetes mellitus," Annals of the New York Academy of Sciences, vol. 967, pp. 363–378, 2002.
117. R. W. Van Der Meer, S. Hammer, H. J. Lamb et al., "Effects of short-term high-fat, high-energy diet on hepatic and myocardial triglyceride content in healthy men," Journal of Clinical Endocrinology and Metabolism, vol. 93, no. 7, pp. 2702–2708, 2008.
118. P. D. Cani and N. M. Delzenne, "The role of the gut microbiota in energy metabolism and metabolic disease," Current Pharmaceutical Design, vol. 15, no. 13, pp. 1546–1558, 2009.
119. M. Diamant, E. E. Blaak, and W. M. de Vos, "Do nutrient-gut-microbiota interactions play a role in human obesity, insulin resistance and type 2 diabetes?" Obesity Reviews, vol. 12, no. 4, pp. 272–281, 2011.
120. M. Daniel, K. G. Rowley, R. Mcdermott, A. Mylvaganam, and K. O'Dea, "Diabetes incidence in an Australian aboriginal population: an 8-year follow-up study," Diabetes Care, vol. 22, no. 12, pp. 1993–1998, 1999.

INDEX

D

E

F

H

I

L

M

N

O

P

R

S

T

U

V

W